CATALOG OF HUMAN SOULS
Book 4

NON-TRADITIONAL PSYCHOANALYSIS
Selected Scientific Articles And Presentations At Conferences

Andrey Davydov and Olga Skorbatyuk

HPA Press

ISBN-10: 0996731237
ISBN-13: 978-0-9967312-3-2

Andrey Davydov and Olga Skorbatyuk

Non-Traditional Psychoanalysis: Selected Scientific Articles And Presentations At Conferences.

Translated by Kate Bazilevsky.

Part 2 Article 1 translated by Anonymous.

Dedicated to the 40th anniversary of the discovery of the Catalog of Human Population by researcher Andrey Davydov.

Table of Contents

PREFACE

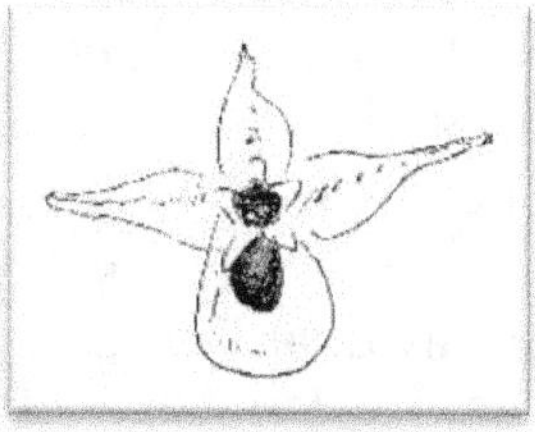

Sketch © 2000 Andrey Davydov

We—authors of the Catalog of Human Souls book series—would like to briefly describe what these books are about right from the beginning because the topic of the Catalog of Human Population is very new for many people.

These books are dedicated to the 40th anniversary from the beginning of research that led to discovery of the Catalog of Human Population. A technology of uncovering of the individual structure of human psyche (or simply put—the soul) was created on the basis of this scientific discovery. The author of scientific discovery of the Catalog of Human Population and this technology is a researcher of ancient books, an expert in Chinese culture Andrey Davydov.

The source of knowledge about the structure of human psyche and the basis for creation of this technology is one of the most ancient and mysterious texts preserved in this civilization—the ancient Chinese monument Shan Hai Jing (translated as the Catalog of Mountains and Seas), which Andrey Davydov managed to decrypt.

After Andrey Davydov created the technology of decryption of Shan Hai Jing (Catalog of Mountains and Seas), it was found that this ancient Chinese monument contains very detailed descriptions of 293 models of human psyche and a lot of other kind of information about the structure of *Homo sapiens*. On this basis, Shan Hai Jing was qualified as the Catalog of Human Population.

In Shan Hai Jing (Catalog Mountains and Seas), the biological type "human" is described as a type divided into 293 subtype structures according to the phenological principle. It turned out that each person, belonging to one of these 293 subtypes from birth, has stable characteristics of this subtype; regardless of race, nationality and particularities of parental psychophysiological structures, which are only minor correctors.

Information about the subtype structure is implanted in the form of a program in the unconscious of a person from birth, and this program determines all of his life: his/her personal qualities and character properties, algorithms of life and functioning, hidden motivational spring, abilities, talents, preferences, inclinations, etc. Natural subtype program is that what is called "psyche", "soul."

Natural program is that individuality, which makes a person different from other people as representatives of other subtype structures of biological type *Homo sapiens.* Each individual "speaks his own language," specified by his subtype program, as the language of values, views, convictions, preferences, which are standard and unchanging for all representatives of a subtype.

In the language of science, a natural subtype program of *Homo sapiens* is called Individual Archetypal Pattern; in simple language—an Individual (Subtype) Program, an Individual Program or a Program.

A human program is recorded in the language of natural images. Images or, using the language of science, archetypes of the unconscious sphere of a person is the language of human "software." The concept of "an archetype" was introduced to psychology by Carl Gustav Jung, but as it turned out, archetypes can be not only of the collective unconscious, but also individual. Therefore, to avoid confusion, in popular texts we prefer to call the language of "software" of *Homo sapiens* by the word "image" instead of "archetype."

Programs of each of 293 human subtypes are recorded by different natural images and a different number of images, meaning that they are endemic, are not similar to one another. *Homo sapiens* is a living system, which, as it turned out, exists and functions strictly on the basis of a natural program implanted from birth, and from this it was concluded that a human is a bio-robot at the genetic level.

The conclusion that "*Homo sapiens* is a bio-robot" is confirmed by that learning someone's natural individual (subtype) program from the Catalog of Mountains and Seas as the Catalog of Human Population, it is possible to find out absolutely everything about this person in great detail; about any aspect of his life and activities, including that what he or she carefully conceals.

In addition, *Homo sapiens*, as a biosystem programmed by nature, has modes of self-regulation and regulation (control from the outside). Modes of regulation are a natural inborn mechanism, just like an individual human program. Their discoverer Andrey Davydov named these modes Individual Manipulation Modes (Manipulation Modes for short). This management tool, which can be applied to any person, also was found in the text of the ancient Chinese monument Shan Hai Jing.

It was found that for every person, as a biosystem, there are three manipulation modes: suppressing, balancing and stimulating. Manipulation modes together with an individual (subtype) program are individual structure of psyche of *Homo sapiens*. Programs and manipulation modes of each subtype differ, from one another. For this reason, people differ from each other by internal characteristics, and individual manipulation scenarios are necessary for each person.

Structure Of Psyche Of *Homo Sapiens*

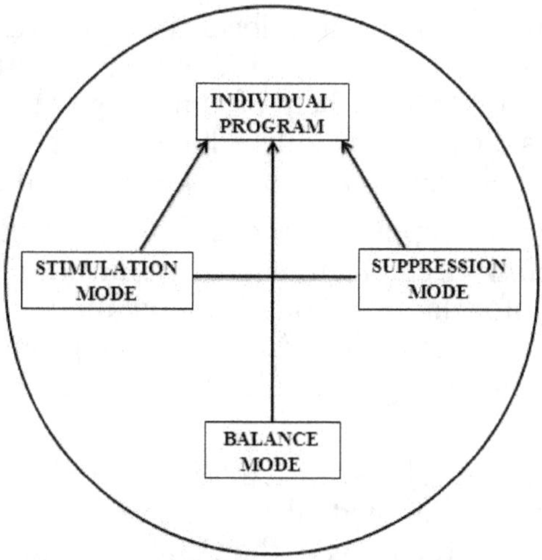

© 2015 Olga Skorbatyuk, Kate Bazilevsky

Individual program, with which a person was born, is the main segment of structure of human psyche (soul). Knowing the natural individual program allows you to find out who he/she (or you personally) is real, without masks, what motives drive this person (or you), and so on.

Suppression mode is a self-control mode, which turns on automatically as a mode of relaxation, pleasure, dream, happy oblivion (a "turn off" button). It is the individual program of people, who belong to a particular subtype. When a manipulator acts out a personality from the suppression mode, it makes an individual experience pleasure, unconsciously get attached to the manipulator (fall in love, etc.), and on this basis obey him/her and fulfill all of his/her requests/demands.

Balance mode is a self-control mode, which turns on automatically when a person needs to become balanced, get in a harmonious, comfortable state without loss of activeness. It is the individual program of people, who belong to a particular subtype. When a manipulator acts out a personality from the balance mode, it makes an individual experience comfort in communication with him/her, trust him/her at the unconscious level and consider him/her the best friend.

Stimulation mode is a self-control mode, which turns on automatically when a person needs to stimulate himself/herself to some actions, become active. It is the individual program of people, who belong to a particular subtype. When a manipulator transmits properties of the stimulation mode in a certain way, it makes an individual experience strong irritation, up to wild rage, and on this basis he/she does that what a manipulator asks for/demands.

Natural manipulation modes described in Shan Hai Jing (Catalog of Mountains and Seas) as the Catalog of Human Population provide a key to managing any person. With the help of knowledge of human manipulation modes, it is possible to factually change a person's physical and psychological state, behavior, reactions in the desired direction.

No one is able to resist application of his personal manipulation modes because it is a natural mechanism, which is built into human psyche from birth. Therefore, information about manipulation modes of a subtype provides unlimited possibilities for influencing any individual as a representative of this subtype and allows self-control and control of any human as a biosystem.

Non-traditional psychoanalysis, a new direction in scientific psychology, allows identifying an individual's subtype structure (psyche) and manipulation modes on the basis of the Catalog of Human Population. It is called non-traditional because it does not use any of the traditional principles, approaches to the study of human psyche, as well as methods (observation, experiment, testing, biographical method, questioning, conversation) since these methods are not needed in order to obtain any kind of information about a person as a stable and identifiable biosystem.

A non-traditional psychoanalyst knows what a person is like, what problems might concern him or her and why, and how these problems can be effectively solved without any kind of contact with this person. All that a non-traditional psychoanalyst needs to know about the subject being studied is the day, the month, the year of his or her birth and the gender. Neither race nor nationality or place of residence of the person being studied matter because the Catalog of Human Population contains descriptions of unchanging characteristics of subtypes of *Homo sapiens* of the entire human population.

For colleagues from the scientific environment we are offering our definition of what human psyche and the Catalog of Human Population are: "*The Catalog of Human Population is a description of a human as a type by subtype structures. Subtype structure ("psyche", "soul") is a combination of individual archetypes, recorded at the genetic level (principle). Expressions and interaction of subtype structures in manipulation modes and phenological algorithms are described with adjustments for gender, age and cultural differences. Information is recorded on six factors.*" This definition was developed by Andrey Davydov—the author of discovery and decryption of the Catalog of Human Population.

It is also necessary to tell a bit about the main source of knowledge (knowledge of that *Homo sapiens* are bio-robots, that human psyche (soul) is "software," that "software" of any person can be easily uncovered, and so on), which was obtained in the course of our scientific research. To date, no one knows for sure who gave and preserved this knowledge, but after it was written down it got the title 山海經 Shan Hai Jing (translated from Chinese as Canon/Catalog of Mountains and Seas).

According to Artem I. Kobzev (a Russian historian of Chinese philosophy, Doctor of Philosophical Sciences, professor, author of over eight hundred scientific papers on the history of Chinese philosophy, science and culture), Shan Hai Jing is an anonymous monument, which presumably dates back to the late III century BC-early II century BC. It consists of eighteen juan (scrolls), combined into two sections: Canon/Book of the Mountains (Shan Jing) or Canon of Five Innermost Mountains (Wu Zang Jing) in five juan and Canon of the Seas (Hai Jing) in thirteen juan.

Legend claims that Shan Hai Jing was engraved on sacred vessels by Bo Yi—an assistant to the wise semi-mythical ruler of the ancient times Yu the Great, who lived in the XXIII century BC. Yu the Great entered the throne in 2205 BC. Authors of the Han epoch attribute authorship of the literary monument Shan Hai Jing to Yu the Great and his companion Bo Yi.

According to legend, Yu the Great dealt with a great flood, which fell upon earth and arranged it. The deedful ruler allegedly came to know its mountains, rivers, their spirits, as well as animals and plants. He ordered his assistant to describe everything that was seen. As a result, these recordings together with images of spirits, fantastic animals, birds and plants were engraved on nine ritual vessels-tripods. Later on, these sacred vessels were lost. However, according to historical annals, prior to their strange disappearance, the text of the Catalog of Mountains and Seas, along with amazing images of representatives of flora and fauna, spirits and deities were copied.

This answers the question why some date Shan Hai Jing back to III-II centuries BC, while others back to XXIII century BC—the Catalog of Mountains and Seas as the source of information appeared during the time of Yu the Great and it got the form of a text much later. We tend to agree with those researchers, who date Shan Hai Jing back to XXIII BC since according to Dan Zhu the Catalog of Mountains and Sea was recorded only after its long-term oral existence (this point of view is expressed in his commentaries to the famous monument Spring and Autumn - Chunqiu). Reports that Yu the Great and his assistant Bo Yi created the Catalog of Mountains and Seas exist in, for example, Wang Chong's (27-97 AD) treatise titled Critical Essays: "When Yu and Bo Yi were taming the waters of the flood—Yu was engaged in calming the water, while Bo Yi in recording information about various "things." And, they created the Catalog of Mountains and Seas."

We will add that apparently in mythologies of almost all cultures of the world (those, which to continue to exist, as well as those, which already sank into oblivion) exist facts, which show that the Catalog of Human Population was present in these cultures. This is not difficult to trace by carefully studying the ancient, archaic cultural layers. However, the source itself was preserved only in one culture—the culture of China. This is not surprising since Chinese culture is not only ancient, but also, in spite of everything, the Chinese manage to preserve it and their traditions from the ancient times to the present day (unlike representatives of other cultures).

This was a brief description of the main points, about which you will be able to learn in detail from the Catalog of Human Souls book series.

*****

In essence, the Catalog of Human Souls book series is one book divided into five parts. All these parts are devoted to a single topic—scientific discovery made by researcher Andrey Davydov, the Catalog of Human Population. Division into five parts was done because materials for this book are quite non-uniform in content and style.

The book Catalog of Human Souls is divided into five parts since fundamental research being carried out by Andrey Davydov and our colleagues is interdisciplinary, and the field of research is wide and multifaceted (psychology, gender relations, sociology, political science, linguistics, sinology and so on), and also because we present results of this research not only in the language of science, but also in the language of popular science.

In addition, we (authors of Catalog of Human Souls books), who are presenting research and discoveries of our laboratory's research supervisor Andrey Davydov, found it necessary to include some information about him as well. It seemed to us that the story of discovery of the Catalog of Human Population might be of interest not only to those, who are already familiar with it, but also to those, who are just learning about the Catalog.

Also, in order not to sound groundless, we decided to offer readers to get acquainted with some of materials from the Catalog of Human Population. These materials are provided in the form of brief demonstration versions from descriptions of natural programs of people. We are of the opinion that presentation of any, even the most captivating scientific theory is, as the saying goes, not worth even a straw, if this theory cannot be tested and used on practice.

Demos that we presented certainly are very short and they are by no means complete descriptions of individuals. However, perhaps they will be quite enough for independent testing, which even a non-professional in the field of psychology will be able to do and get a confirmation that the Catalog of Human Souls (a nonscientific title of the Catalog of Human Population) really exists. And, this means that now he or she can use this Catalog in daily life.

We—authors of the Catalog of Human Souls book series—have the honor to not only tell about some of the research done by our laboratory's research supervisor Andrey Davydov and our colleagues, but also to participate in the development of the Catalog of Human Population. However, it must be stated right away that in all sections where we tell about the technology, methodology of non-traditional psychoanalysis—we are only narrators. Technology of decryption of the ancient Chinese monument Shan Hai Jing was created over 20 years ago and its authorship belongs to Andrey Davydov.

*****

The Catalog of Human Souls book series contains the following five books:

Book 1. *Homo Sapiens* Are Bio-Robots. Human "Software" (by Olga Skorbatyuk and Kate Bazilevsky). This book tells about the new method of obtaining information from the unconscious sphere of a human and about the Catalog of Human Population in the form of answers to the most frequently asked questions. In this book, we also presented three hundred selected research topics, which were developed in our laboratory between 1974 and 2014.

Book 2. Hack Anyone's Soul. 100 Demos Of Human Programs From The Catalog Of Human Population (by Olga Skorbatyuk and Kate Bazilevsky). In this book, we are offering one hundred brief demonstration versions taken from descriptions of human programs in the Catalog of Human Population. Of course, these descriptions are only a tiny fraction of information that the Catalog contains about each person.

However, we think that this is enough for independent testing and to obtain confirmation of existence of the Catalog of Human Population.

Book 3. Human Manipulation Modes. Either You Are Manipulating Or You Are Being Manipulated (by Olga Skorbatyuk and Kate Bazilevsky). This book is entirely devoted to manipulation modes—the natural toolkit for controlling a system called "a human," which was discovered in the ancient Chinese monument Shan Hai Jing (which turned out to be the Catalog of Human Population). In this book, we explain what individual manipulation modes are and how to use them to manipulate people. After all, with the discovery of the Catalog of Human Population, only two positions in society are left—either you are manipulating or you are being manipulated. And, as proof of that this is really so, in this book we provided the scenario of suppression mode for manipulation of people, who were born on October 12th of leap years or October 13th of common years. Despite that this is only one of four modes of manipulation of those people, who were born on the dates indicated—it will be sufficient for testing.

Book 4. Non-Traditional Psychoanalysis. Selected Scientific Articles And Presentations at Conferences (by Andrey Davydov and Olga Skorbatyuk). This book presents some of the scientific articles and presentations at scientific conferences done by Andrey Davydov, the author of scientific discovery of the Catalog of Human Population. It also includes several scientific articles (which are also chapters of the textbook titled Archetypal Pattern. Fundamentals of Non-Traditional Psychoanalysis.), authors of which are founders of non-traditional psychoanalysis: Andrey Davydov and his colleague—psychologist Olga Skorbatyuk.

Book 5. Shan Hai Jing—A Book Covered With Blood. The Story Of Developers of the Catalog of Human Population (by Kate Bazilevsky). This book uncovers the story of developers of the Catalog of Human Population—Andrey Davydov and Olga Skorbatyuk— that is related to obtaining of political asylum in the United States of America due to many years of persecution by the Federal Security Service of Russian Federation (FSB, former KGB) and attempts to kill them in order to take possession of their research product—the Catalog of Human Population.

*****

From the beginning, as authors of the book series Catalog of Human Souls, we would like to apologize for the lack of a single language style and popular language. Our colleagues from the scientific environment, professional psychologists or those, who are interested in scientific psychology, can satisfy their curiosity about the theory that lies at the basis of technology that we use to identify the individual structure of psyche in

Book 4 of this series (Non-Traditional Psychoanalysis. Selected Scientific Articles And Presentations At Conferences.).

The other four books in this series are intended for the widest audience. From our point of view, any person, regardless of education level, must have an opportunity to get acquainted with results of our research. After all, Shan Hai Jing, as the Catalog of Human Population, was left to all humanity. Therefore, in many of our books (despite that they are all devoted exclusively to scientific topics) we try to use the literary language, avoiding frequent use of specific terminology. However, we were unable not to use jargon at all, and for this we beg your pardon. We are not writers, our main work is scientific research, and in books we are sharing with the audience some of the results of this activity, according to the common practice in the scientific community.

Speaking of language that we use to describe the topic of our research... It was noticed that some people are perplexed by that we use the word "soul," as it is not customary to use this word in the scientific community. However, as it was found, human soul, which is the same as psyche and subtype structure—exists. And, it exists regardless of someone's opinions about it.

Soul (psyche), as "software" of *Homo sapiens* is not a psychological, religious or metaphysical value—as it turned out, it is purely a natural value. Therefore, like any natural phenomenon (rain, wind, electricity, gravity), the soul functions regardless of whether people know about it or not, and regardless of what they think about this. People can argue as much as they want whether soul exists or not, and fantasize as much as they want about what it is, but the natural mechanism called "soul" will continue to work. In every person, while he is alive.

Therefore, every person has the right to continue debating whether or not the soul exists and fantasize about what it is, but we prefer to study the human soul as an existing phenomenon (which, by the way, is possible to see, hear, smell and touch, if you have information about individual structures of psyche) and use results of this research on practice.

We consider the question of what words to use to call the natural phenomenon "soul" unimportant when it is possible to study it as a phenomenon instead of talking about it. We use the word "soul" because in any language it accurately reflects the essence of the phenomenon—that foundation, on the basis of which any human being lives, and because of which he is alive; that what is called "closer than the body." And, unlike our colleagues-psychologists, we are not forgetting that "psychology" translated from the ancient Greek means "science about the soul" (from the ancient Greek ψυχή - "soul," λόγος - "teaching").

*****

Now, let's get back to the Catalog of Human Souls books series. Those people, who are already familiar with our research and our other books, usually have questions, to some of which we would like to provide answers in the Catalog of Human Souls book series. For example, people are interested to know who exactly is working on compiling of the Catalog of Human Population, and why it is not possible to find information anywhere, including on the Internet, about all members of our laboratory's staff. Also, people want to know how they can get information from the Catalog of Human Population about themselves or other people.

With regard to the question about developers of the Catalog of Human Population and other staff members of our laboratory, it really is difficult to find reliable information about them and here is why. This happens for two reasons: the first reason is explained using an example with the Internet in the Introduction to Book 1 of this series titled Homo sapiens Are Bio-Robots, and the second reason is confidentiality of all of this kind of information and is related to security. The number people in our laboratory, who work on compiling of the Catalog of Human Population, is not three people, as it might seem. However, unfortunately, we cannot provide any information about who these people are. Reasons for this are detailed in Book 5 of this series (Shan Hai Jing—A Book Covered With Blood. The Story Of Developers Of The Catalog Of Human Population).

Due to persecution because of the main subject of our scientific research—the Catalog of Human Population—by a group of employees of Federal Security Service of Russian Federation (FSB, former KGB), headed by colonel of Foreign Intelligence Service (data for the period 2000-2004) Andrey Dmitrievich Polonchuk, we do not have the right to provide any kind of information about our laboratory's staff. You can get acquainted with the evidence of these prosecutions in Book 5 mentioned above. Originals of documents presented there are stored in the United States Department of Homeland Security, Federal Bureau of Investigation (FBI), etc.

Therefore, if anywhere in public sources, for example on the Internet, you come across a statement that someone is an employee of the Special Scientific Info-Analytical Laboratory—Catalog Of Human Souls or the Human Population Academy, and is engaged in developments related to decryption of Shan Hai Jing and compilation of the Catalog of Human Population—it can be either disinformation or fraud, or both. However, we do have partners and affiliates. Therefore, it is always best to check with us.

Beware that some research works of Andrey Davydov (the author of discovery of the Catalog of Human Population) and the Special Scientific Info-Analytical Laboratory—Catalog Of Human Souls led by him were stolen by a group of employees of the Federal Security Service of Russia (formerly KGB) headed by Andrey Dmitrievich Polonchuk—colonel of SVR FSB of Russian Federation (information about his rank refers to 2000-2004). Therefore, it is possible to find offers on the Internet (in different languages) to purchase information from the Catalog of Human Population, even though this information is in no way related to the Catalog or to natural human "software."

Such offers can be found at the following websites: in Russian – Mountains and Seas of Self-Knowledge http://mountaseas.com, Catalog of Human Population (Catalog of Mountains and Seas) http://vk.com/chp_lab, Find The Answer Within Yourself http://www.facebook.com/groups/mountaseas.2905/, Catalog of Human Population http://vk.com/googlite; in English – Shan Hai Jing Lab/Shan Hai Jing Laboratory http://www.facebook.com/SHJLab/info, http://twitter.com/SHJLab and http://shjlab.wordpress.com; etc.

It is extremely dangerous to buy products offered by fraudsters. Since some time back Andrey Davydov suspected that sooner or later, in one form or another piracy will occur and made some inaccuracies in materials related to Shan Hai Jing; without taking them into account it is impossible to correctly put together the base material.

Perhaps, after realizing this A. D. Polonchuk's group decided to get at least some, at least material benefits from stolen works and this is how a strange, ugly product was born, which has nothing to do with information from the Catalog of Human Population as the decrypted ancient Chinese monument Shan Hai Jing.

The consequence of this is that those, who get information from fraudsters cause irreparable damage to their health; both physical and psychical. Unfortunately, our laboratory already has a significant amount of information about cases when people, who requested information from the Catalog of Human Population from those, who officially have nothing to do with this research, experienced serious damage in various matters and of varying degree of severity. The reason for this is the fact that human psychophysiology functions on the basis of images. Therefore, if a person receives images and decryptions (which essentially are images as well), which are in no way related to his/her personal psychophysiology and begins to use this information on practice, then consequences might be as follows:

- psychical disorders of any kind, which cannot be cured
- somatic disorders of any kind, which cannot be cured
- death, as a result of psychical and physiological disorders

Therefore, please be careful. You have every right to turn to anyone for information from the Catalog, but we issued this warning and we will not be held responsible in any way for possible consequences of your decisions.

You can get information from the Catalog of Human Population about yourself or any other person directly from our laboratory right now. For convenience of our clients, the system of issuance of any kind of information directly from developers of the Catalog of Human Population is made in such a way that they do not need to contact us. Anyone can view prices and pay for an order here - https://www.humanpopulationacademy.org/pricing/.

And, for those people, who would like to not just get materials from the Catalog of Human Population sold on our website, but also contact us (for example, with a business proposal, to order of some specific services, to get consultations, to become a student, etc.)—please refer to Human Population Academy's Contacts page at https://www.humanpopulationacademy.org/breakthrough-discovery/contacts/.

Purchasing information from the Catalog through our website or by contacting us are the only ways to get information from the Catalog of Human Population directly from its developer without the risk of coming across fraudsters.

It should be noted that no one ever persecuted our clients; they never were and are not in any danger because those who persecuted and continue to persecute us were never interested in our clients as they do not have information about the technology of decryption of Shan Hai Jing. In addition, we always respect confidentiality of any person, who turns to us for information from the Catalog and we never share information about our clients with anyone.

*****

In conclusion, we would like to state the principled position that we adhere to as authors of books (and, currently there are over three hundred of them)—we do not in any way claim to be gurus; we are more comfortable with the position of students. However, due to the fact that we engaged and continue to engage in research work, we are able to obtain information that can be interesting and useful not only to us. And, we are just legalizing this information for those, who prefer not to believe, but to know. To know how the world works and what is his or her own, personal structure.

Another principal point is that in our books, we do not share our personal opinions, hypotheses. Everything that we state is information from the ancient sources that we are studying and facts obtained during the course of our research activities, which have been tested, as it is customary in scientific practice. We prefer to keep to ourselves our opinions and fantasies, as insignificant in comparison with that what is stated in our books.

Also, on behalf of our colleagues from the laboratory and ourselves, in celebration of the 40th anniversary since the beginning of research of the ancient Chinese monument Shan Hai Jing as the Catalog of Human Population, we would like to express deep gratitude to all of our relatives, friends, acquaintances and colleagues from the scientific environment, as well as to all others, whom we happened to come across in life, for that they failed to prevent us from carrying out this research.

Olga Skorbatyuk and Kate Bazilevsky
Los Angeles, California
June 2015

INTRODUCTION

THE ANCIENT CHINESE MANUSCRIPT SHAN HAI JING TURNED OUT TO BE THE CATALOG OF HUMAN POPULATION

Olga Skorbatyuk and Kate Bazilevsky

"Another 500 li to the east, there is a mountain called Rooster-of-the-year. At the top there is a lot of gold, at the bottom—a lot of cinnabar (Danhu). A black river flows out of there to the south and empties into the sea. Zhuan fish exists there, it is similar to a frog (Fu), but with a bristle. It emits a sound like a grunting pig..."

Shan Hai Jing (Catalog of Mountains and Seas)

Catalog of the Southern Mountains

3rd Chapter

Neither ancient nor modern sages, nobody knows for certain what It is, who gave and preserved this knowledge. However, after it was written down, it was named Shan Hai Jing—The Catalog of Mountains and Seas.

Attribution of authorship, which is required by the ancient Chinese tradition, also occurred. Bo Yi was credited as the creator of this significant work, which contains mysterious knowledge. He was an assistant to Yu the Great—the mythical ruler of China, who, according to legends, lived sometime during the XXI century BC. The legends claim that Yu the Great dealt with a great flood, which fell upon earth and arranged it.

The deedful ruler allegedly came to know its mountains, rivers, their spirits, as well as animals and plants. It was he, who ordered his assistant to describe everything that was seen. As a result, these recordings together with images of spirits, fantastic animals, birds and plants were engraved on nine ritual vessels-tripods. Later on, these sacred vessels were lost.

However, according to historical annals, prior to their strange disappearance, the text of the Catalog of Mountains and Seas, along with amazing images of representatives of flora and fauna, spirits and deities were copied.

For all of his deeds, Yu—the legendary hero of the prehistoric Celestial Empire—was proclaimed the Great and many centuries later was canonized by Confucians, whose religion reigned in China until the middle of the XX century. However, the work, which he ordered to create, was out of luck. Shan Hai Jing was not included in the category of Confucian classical books, such as Four Books and Six Classics, which contain sayings of Confucius himself, his followers or works collected and edited by the Teacher. It is unlikely that this was accidental.

Ancient and medieval commentators do not correlate the Catalog of Mountains and Seas with classical heritage, thinking that it has nothing to do with philosopho-ethical teachings of Confucius or a number of other schools of philosophy. Shan Hai Jing was considered a poorly understood book on historical geography, flora and fauna of prehistoric China. In their comments to this treatise, some researchers tried to determine the geographical location of natural objects describes in it and classify fantastic flora and fauna. However, their attempts did not result in anything.

In the end, it was decided that the Catalog of Mountains and Seas is a mythological work of the most ancient Chinese people, which was not included in canonical books due to its incomprehensibility.

Indeed, the Catalog describes a strange world, which even the keenest contemporary fiction writer would not be able to think-up. Outlandish animals, reptiles, some unrealistic birds—in short, a figment of imagination of the ancients, who, at first glance, created that world on purpose, as if for fun and to ridicule the now living: go ahead, scratch your heads over what this means. However, it is unlikely that the ancients would have engaged in such an unserious undertaking as creation of images of fantastic animals that never existed for us, even as myths. This would have required too much work and mental efforts from people, who had to get food by hard labor and in harsh conditions day and night.

"...The river of evaporations begins there, flows to the west and flows into the river Tan. There is an animal there similar to a mouse, but with head of a rabbit and body of a moose. His cry is similar to howling of a dog..." It is quite difficult to imagine such a monster. Or, for example: an animal, which resembles a rabbit, but has a bird's beak, eyes of an owl and tail of a snake. There are so-called myths, mythical images, which are simply listed in Shan Hai Jing. Although it is necessary to note that until present time real, in our understanding, myths with action and storyline have survived in China, as well as in other nations of the world. Fairy tales and literary works were created on their basis.

Thus, the world described in Shan Hai Jing is not clearly understood, even when considering it as a purely mythological work. Modern scientists, who are involved in research of this fantastical monument, note that many of the names are found only in the Catalog and their true meanings are lost forever. Other names can be found in ancient dictionaries, but synonyms by which they are revealed, for the most part, are also unclear.

Nevertheless, what is Shan Hai Jing? Is it a list of representatives of flora and fauna, data on historical geography, which absolutely no one can understand, or is it something else? It turned out that it is something else. And, this "something else" correlates with the structure of human psyche very well.

Currently, a hypothesis exists that humans obtained some "knowledge" from extraterrestrial beings. According to some sinologists, there is even evidence in China of that extraterrestrials visited Earth. The mythical Yellow Emperor himself, who is considered the progenitor of all Chinese people, supposedly was an "alien" as well. Upon expiry of his term on Earth, he returned to the Yellow star and, most likely, left "cosmic knowledge." And, it is possible that among them he also left knowledge about identifying structure of human psyche.

It is possible that at the time they were recorded using some mathematical or other methods, which in the end were lost because eventually they became incomprehensible for following generations. However, they there recorded in consciousness of the initiated in the form of comparative images of fantastical animals, amazing trees, stones, and this was reflected in Shan Hai Jing. In any case, Shan Hai Jing is perfectly suitable for identifying the structure of human psyche.

It is unlikely that someone will deny the existence of specific types of geographical regions, water flows, reservoirs, and elevations. They all have their own inhabitants, including people, of course. All this "cooks" in one natural "pot" of a given region, where each animal, plant, human are ingredients of a natural "brew." Being born in a certain geographical "pot," every person soaks in the natural surroundings of his or her environment—animals, birds, reptiles, plants— genetically and psychologically. It is necessary to note that this process occurs non-uniformly. One person's psychophysiology soaked in functional abilities of a bull's body and a rooster's head; another person's—elements of other animals. Aren't similar monsters described in the Catalog of Mountains and Seas?

Form as a whole, as well as separate parts of a human are peculiar "conductors," which manifest elements of his psyche. In time it forces "conductors" to acquire characteristics that correspond to psychical analogues or their segments in nature; in other words—representatives of the surrounding flora and fauna. Therefore, human psyche well can be similar to a peculiar conglomerate of natural analogues or their parts.

By the way of selection according to a certain angle, scale and outline, it is possible to uncover to what segment of a specific animal a particular part of the human body belongs. Based on practice described in the I Ching (Book of Changes), which also has the most ancient and vague origin and has not been unraveled up to now, where each line of a hexagram correlates to segments of a human body and an animal—makes it possible to get a quite exact structure of psyche (sub-consciousness and unconsciousness) of any person being studied with the same functional diapason as that of an animal.

Hexagrams are certain graphical symbols, which, according to some scientists, record a universal set of inborn psychical structures, or, as they are also called, "archetypes."

The next stage is identification of the resultant structure by mythical, from the point of view of a modern person, representatives of flora and fauna in Shan Hai Jing. Knowing the date (day, month, year) of birth of a person, it is easy to find his or her description in the Catalog of Mountains and Seas, which begins with the Catalog of the Mountains according to cardinal points, where each mountain has its own range of habitation of amazing animals, plants and minerals.

Naturally, this method of determining psyche using Shan Hai Jing can be applied not only to an individual, but also practically to any nation as a whole. This requires only a small correction. Each area, region is characterized by its own ecosystem, biorhythm, and energy. In addition, in order to compare the similarity between a part of a human and animals (plants), it is necessary to consider representatives of fauna, flora and folklore, which are native to a person being studied.

Thus, as already mentioned above, Shan Hai Jing, begins with the Catalog of the Mountains according to cardinal points: Southern, Western, Northern, Eastern and Central. They symbolize individuals, and rivers are their lives, which at one moment flow into a sea, at another into a lake or a swamp.

The ancient Chinese teaching about the five elements also makes it possible to determine to a great extent very fine details of psychological structure of a human. For example, water corresponds to the north, its color is black, and emotions—fear. By the way, the first emperor of the unified China (III century BC) Qin Shi Huang, who was born under the element of water, wore black clothing and remained in history of the Middle Kingdom as the cruelest despot. He was feared so much that when he died, people were afraid to come near him for a few days...

When studying the remaining sections of Shan Hai Jing (Catalog of Regions Beyond the Seas, etc.), it becomes clear that the Catalog of the Mountains describes psychological structure of a human, who has not yet advanced in his development from the world of animals and must "work through" the program that was implanted in him or her by nature.

A more perfect stage, liberation of a human from previous psychological structures and attainment of knowledge of the highest level (rational), can be traced in the next sections of Shan Hai Jing. There, for the most part, deities prevail among the inhabitants. Naturally, in order to determine the way of transition to a more perfect structure of psyche—a more complex technology is necessary, which is based on the most ancient knowledge stored in Shan Hai Jing, I Ching, and the Teaching of the five elements.

*****

A reader might have doubts: is this possible? As it turned out—this is possible. It turned out that the ancient Chinese were not so naïve as to pass on worthless knowledge to future generations. They conveyed, passed on to us information about perfect humans, who comprehended the connection between their internal world and the macrocosmos, who were capable of understanding the essence of changes in nature, flying on the clouds and transmitting information from a distance. However, from our point of view, the most important thing is that a text titled Shan Hai Jing (Catalog of Mountains and Seas), which turned out to be the Catalog of Human Population, was left for us.

The ancient Chinese manuscript Shan Hai Jing was qualified as the Catalog of Human Population on the basis that the author of this scientific discovery—Russian researcher-sinologist Andrey Davydov found in this source detailed descriptions of subtypes of the biological type *Homo sapiens*. It was found that by nature the biological type "human" is divided into 293 subtypes, representatives of which, despite principal physiological similarity, have models of structure of psyche that principally differ from each other.

It turned out that a human is a bio-robot because he lives and functions solely on the basis of a strict program, just like all the other natural objects on this planet live and function: flora and fauna, rocks and minerals, mountains and rivers, the planet Earth itself and other planets of the Solar System. One of 293 natural programs turn on at the moment of birth[1] in each representative of the biological type *Homo sapiens*, and a person begins to live and function strictly in accordance with this natural program, regardless of whether he knows about its existence or not.

Natural human program, as it turned out, is that "soul," which psychologists, anthropologists, sociologists, philosophers, esotericists, religions of all confessions, etc. have been and still are looking for without success.

Every person being born with a natural program, which is, using technical terms, "inbuilt" in his psyche, is incapable of going beyond this program during the course of his or her life. It is this natural arrangement of *Homo sapiens*, which allows us (researchers of Shan Hai Jing, who have special technology of decryption, unlike other researchers of this ancient book) to easily uncover the content of the unconscious sphere of any individual. And, from there we get any kind of information about this person: his or her true, natural qualities of personality, motives, life algorithms, goals and objectives, and much more.

On the basis of this most ancient of texts that survived in this civilization—the Catalog of the Mountains and Seas—it became possible to compile the Catalog of Human Population. This Catalog is being created by its developers in the form of 293 descriptions of *Homo sapiens* subtype structures according to 6 factors: intellectual, physical, nutritional, emotional, sexual, and environmental. The unique system of decryption of this ancient source and the 6-factors system of descriptions of psychophysiological structure of a human was created by Andrey Davydov in the 80s of the XX century.

In conclusion it is possible to say that due to ancestral heritage, which survived only in one of the existing cultures—the Chinese culture—in the form of one of the most ancient and mysterious manuscripts, Shan Hai Jing, as well as due to scientific approach of researcher-sinologist Andrey Davydov to sources that are traditionally considered esoteric—psychology finally got every right to be called a science[2], and humanity got the Catalog of Human Population. Of course, it is a pity that this Catalog, which was discovered 40 years ago and is actively being used for over 20 years, only now, due to some causes (which are described in detail in other sources[3]), is becoming available to people around the world[4]. However, we are convinced that this is exactly how it should be, as the Catalog of Human Population describes *Homo sapiens* regardless of their racial, national and other differences, and, therefore, it should belong to all humanity, and not just any one, very small part of it.

Bibliography

Yanshina, E. M. (Trans.). (1977). *Katalog Gor I Morej (Shan Hai Tszin)* [Catalog of Mountains and Seas (Shan Hai Jing)]. Moscow: Nauka.

Jung, C. G. (1991). A. M. Rutkevich (Ed.). *Arkhetip i simvol* [Archetype And Symbol]. Moscow: Renessans.

Lukyanov, A. E. (1994). *Nachalo drevnekitayskoy filosofii: "I tszin", "Dao de tszin", "Lun' yuy"* [The Beginning Of The Ancient Chinese Philosophy: I Ching, Tao Te Ching, Lun Yu]. Moscow: Radiks. ISBN 9785864630266.

[Available at multiple libraries http://www.worldcat.org/title/nachalo-drevnekitaiskoi-filosofii-i-tszin-dao-de-tszin-lun-iui/.]

Otyugov, A. A. (1998). Corr. of RAO L. Verbitskaya & Assoc. Prof. B. Sokolova (Eds.), *Pervyy Rossiyskiy Filosofskiy Kongress: Chelovek – Filosofiya – Gumanizm* [First Russian Philosophical Congress. Human Being – Philosophy – Humanism]: *Vol. 7 Philosophy and Human Problem: Psychological Nature Of Mythological Consciousness.* St. Petersburg: SPSU Publishing House. ISBN 9785288018947. Presentation at the First Russian Philosophical Congress: Human Being – Philosophy – Humanism (in Russian), St. Petersburg. Retrieved from http://mirlyn.lib.umich.edu/Record/003947324 B4231.R6751997 (subscription required). [Also available at other libraries http://www.worldcat.org/title/chelovek-filosofiia-gumanizm-pervyi-rossiiskii-filosofskii-kongress/.]

Zarochentsev, K. D., & Khudyakov, A.I. (1998). Corr. of RAO L. Verbitskaya & Assoc. Prof. B. Sokolova (Eds.), *Pervyy Rossiyskiy Filosofskiy Kongress: Chelovek – Filosofiya – Gumanizm* [First Russian Philosophical Congress. Human Being – Philosophy – Humanism]: *Vol. 7 Philosophy and Human Problem: The Crisis Of Modern Psychodiagnostics.* St. Petersburg: SPSU Publishing House. ISBN 9785288018947. Presentation at the First Russian Philosophical Congress: Human Being – Philosophy – Humanism (in Russian), St. Petersburg. Retrieved from http://mirlyn.lib.umich.edu/Record/003947324 B4231.R6751997 (subscription required). [Also available at other libraries http://www.worldcat.org/title/chelovek-filosofiia-gumanizm-pervyi-rossiiskii-filosofskii-kongress/.]

Davydov, A., & Fedoruk, V. (1999). *Shan Khay Tszin: Mify Ili Struktura Psikhiki?* [Shan Hai Jing: Myths Or Structure Of Psyche?]. Moscow: *Power Of Spirit*, 32-35.

Ezhov, V. V. (2003). *Mify Drevnego Kitaya* [Myths of Ancient China]. (Ill. *Myths of Nations of the World*) Moscow: Astrel.

Davydov, A., & Skorbatyuk, O. (2014). K. Bazilevsky (Trans.). *IDEOLOGY OF RELIGIONS. Scientific Proof Of Existence Of "God": The Catalog Of Human Population.* (Original work published 2014 in Russian, ISBN 9781311946690.). San Diego, CA: HPA Press. ISBN 9781311413932 ISBN 9780988648593

Footnotes

[1] To be more specific—at the moment of cutting of the umbilical cord that connects a baby to the mother during pregnancy.

[2] The subject of psychology is human psyche, or in simple terms—"soul" (from the ancient Greek ψυχή - "soul"; λόγος - "teaching"). However, prior to the discovery of the Catalog of Human Population by researcher Andrey Davydov, psychology did not have a clear, precise, unambiguous and comprehensive answer to the question "What is the soul of a human, and what is its structure?"

[3] Refer to: the official website of the Special Scientific Info-Analytical Laboratory—Catalog Of Human Souls and the Human Population Academy (http://www.humanpopulationacademy.org); book by Kate Bazilevsky titled *Shan Hai Jing—A Book Covered With Blood: The Story Of Developers Of The Catalog Of Human Population*; book by Olga Skorbatyuk and Kate Bazilevsky titled *Homo Sapiens Are Bio-Robots: Human "Software."*

[4] Andrey Davydov's research began in 1974 in Russia (USSR at the time), and already in the early 90s the technology, which he created for decryption of Shan Hai Jing in order to obtain descriptions of human subtypes from it, was verified in Russian scientific institutions. From that moment 20 more years passed until this technology started being actively used in practice, mainly in the Russian Federation (former USSR).

PART 1

PRESENTATIONS AT CONFERENCES

PRESENTATION 1

PRESENTATION (SAINT-PETERSBURG (RUSSIA), 1997) AT THE FIRST RUSSIAN PHILOSOPHICAL CONGRESS: HUMAN BEING – PHILOSOPHY – HUMANISM. PHILOSOPHY AND HUMAN PROBLEM. *IS SHAN HAI JING THE ORIGINAL CATALOG OF PSYCHOPHYSIOLOGICAL HUMAN STRUCTURE?*

Vladimir Fedoruk and Andrey Davydov

Edited by Corr. of RAO L. Verbitskaya & Assoc. Prof. B. Sokolova.

Translated by Kate Bazilevsky.

1. Problems with identifying the structure of unconscious still exist. This can be called attempts to identify the foundation on the basis of consequence, i.e. to clarify the essence of the issue on the basis of facts that already occurred. Often due to lack of clear objective criteria, experts in the field of psychology cannot clearly describe human psyche and its manifestations in a particular situation. The main subject of psychology is a human, who is being observed and experimented on, and he is the main value rather than the derived or expected result.

2. Turning to the problem of identifying human psychophysical structure, we were prompted to study the ancient Chinese work titled Shan Hai Jing (SHJ), and in particular its first part Shan Jing (SJ), contents of which bring scientists to an impasse and so they attributed it to myths. Meanwhile, it appears that SJ and partially I Ching are perfectly suitable for identifying human psychophysical structure and contain quite complex technology of defining it. The difficulty is in that knowledge that lies in various scientific fields—such as biology, mineralogy, medicine, geography, etc.—is required in order to process the information. Already based on its

title, the ancient book SHJ is a geographical catalog. Therefore, it is logical to begin with consideration of this topic. It is claimed that every biocoenosis is connected to a region that it occupies. Geographers have established concepts: a mainland, an island, a flatland, etc. All of them have naturally defined boundaries. Then, there are smaller magnitudes: small rivers, small hills. They are also natural boundaries of natural holding capacities. Each of these holding capacities has its own environment and inhabitants, including humans. It is a particular geographical nesting doll. This scheme is used in SHJ as well. Fantastic creatures inhabit "holding capacities" that are described in the catalog. Ancient Chinese dictionaries provide explanations for them, but synonymous, by which they are revealed, are often unclear. All of the mythical representatives of flora and fauna live and "brew" in these "holding capacities," as in a pot. There is a lot suggested to eat or try for the purpose of treatment, as well as for nutritional or prophylactic purposes. Each of the ingredients separately has its own taste, its own composition. However, once they are in the "brew," they mutually soak in taste, smell. A simplest excursus allows to suggest that people, who are born within certain geographical boundaries, as in "holding capacities-pots," where they are one of ingredients—"soak in" the given ecological system.

Form as a whole, as well as separate parts of a human are "conductors" of his psyche, which in time acquire characteristics that correspond to psychical analogues. The latter are a set of various forms that exist in the natural environment. This set is determined by way of identification of the form as a whole, as well as separate parts of a human with forms of life on earth or with separate forms of life selected by likeness. Human psyche as a whole is like a peculiar conglomerate, which consists of separate likened forms and has the same diapason of activity as a natural analogue or part of an analogue. Uncovered diapason must be corrected. Correctors are a geographical region and a number of other parameters.

3. The main reference points in SJ are mountains. They have specific names and are positioned, oriented towards each other. Each mountain has its own microworld with universal and peculiar magnitudes, while remaining the main magnitude in the environment that it constitutes. All magnitudes are etalons of measuring a species. "Mountains" are stability of existence and a worthy magnitude of form by its significance in the geographical sense, which is difficult to substitute with something more grandiose. As for religious, mythological interpretations of the image "mountain"—here it is possible to find enough interpretations, which allow to put an equal sign between the words "mountain" and "species." In this case, "species" or "mountain" is a psychical biotype.

Flora described in SJ has the same direction, but only in the form of a cipher of functionality, which needs to be translated into inherent abilities of species that is indicated by a mountain in SJ. Each plant can exist in certain specific conditions. With their help it is possible to decrypt and get an understanding of aspirations, intellectual constructions of a human. If there is no possibility to identify the flora, then functionality is identified through the original form, structure, color, myth, taking into account national particularities. Those parameters of species, by which it will consistently function are identified through form.

Rivers are a required attribute of SJ. The mythological interpretation allows to determine the purpose of the image river, its functional particularities. Combining natural parameters with functionality of an image through the prism of a myth, depending on nationality, provides character of life of a species' representative.

A variety of minerals and metals are mentioned in SJ. It is possible to draw a conclusion about the diapason of functions by researching the diapason of properties of a mineral. Each mineral has its own individual characteristics, taking into account time, temperature, etc. Location of a mineral in SJ determines spiritual functions of a human, possibilities in time and space. The same can be stated about the fauna described in SJ.

4. By way of selection by likeness, which was already mentioned, it is possible to identify to which specific segment of a specific animal a certain segment of the human body corresponds. Comparing must be done according to configuration (contour) on the same scale and angle. Using such method and basing on practice described in I Ching, where each line of the hexagram correlates a segment of the human body and an animal, and the result is the exact structure of psyche (unconscious) of any person under review with the same diapason of activities as that of an animal. The next stage is collation of the derived structure with a description in SJ. Knowing the date of birth and using the eastern lunar calendar, it is possible to find the section SJ, to which a person under review pertains. Using the technology described above, it will be possible to determine exactly what the boundaries of psychophysical abilities of a person under review are, while applying certain correctors.

5. One of them is the geographical location of birth and period of residence there until puberty. Geographical space is naturally divided into regions with an areal of inhabitation by various biological types. To a large extent, habitat conditions determine the phenotype of the species, and a human is not an exception. Having grown up in a particular ecosystem, he carries all changes of this system.

6. Another corrector is mythopoetic structure of nationality, to which a person under review belongs. Thus, SHJ (and, in particular, SJ) can be considered as the catalog of psychophysical structure of both an individual and certain ethnos.

References

Davydov, A., & corr. of ITAR-TASS Fedoruk, V. (1998). Corr. of RAO L. Verbitskaya & Assoc. Prof. B. Sokolova (Eds.), *Pervyy Rossiyskiy Filosofskiy Kongress: Chelovek – Filosofiya – Gumanizm* [First Russian Philosophical Congress. Human Being – Philosophy – Humanism]: *Vol. 7 Philosophy and Human Problem: Is Shan Hai Jing The Original Catalog Of Psychophysiological Human Structure?* St. Petersburg: Saint Petersburg State University Publishing House, 7, 355-357. ISBN 9785288018947. Presentation at the First Russian Philosophical Congress: Human Being – Philosophy – Humanism (in Russian), St. Petersburg (1997). [Available at: University of Michigan http://mirlyn.lib.umich.edu/Record/003947324 B4231.R6751997 (subscription required), St. Petersburg's Central City Public Library Named After V. V. Mayakovsky http://www.pl.spb.ru/structure/zali/ ББК87Ч-391 (subscription required), and multiple other libraries http://www.worldcat.org/title/chelovek-filosofiia-gumanizm-pervyi-rossiiskii-filosofskii-kongress/]

PRESENTATION 2

PRESENTATION (MOSCOW (RUSSIA), 2002) AT THE INTERNATIONAL CONFERENCE ON PROSPECTS OF PRESERVATION AND DEVELOPMENT OF UNIFIED CIVILIZATION OF THE PLANET: CULTURE, ECOLOGY, COSMOS. *CATALOG OF HUMAN POPULATION*.

Andrey Davydov

Translated by Kate Bazilevsky.

This report is devoted to the topic of Systemic Research in the Field of Human Psychophysiology. In this topic, the following question was interesting to me: why with principled psychophysiological sameness of humans, qualities differ? It can be noted that processes of describing, explaining, prognosticating are being continued, and in this report I will try to briefly describe my attempt to make some contribution to this fundamental research. I share the view, in which the search for answers to questions that are part of this topic is simultaneously being undertaken in different fields and is not limited to some separate branch of science.

From the beginning, the main interest was and is the method of managing the behavior of a system and management of artificially created systems of needed behavior. A system stands for the human species in general and a specific person in particular. Information in this field allows speaking about form and the functional diapason of form in the environment, but not about consistent rules of characteristics of the functional diapason. Here, a form stands for human psychophysiology. Research in the field of physiology, in my view, is very much ahead of achievements in the field of psychology. The latter, as it is admitted by some psychologists, accumulated great experimental material, developed unique methods of experimental study of psyche; however, today psychology is at an impasse. It lost the very subject of its application— determination of its place in the solution of the main problem of psychology: what is psyche.

Initially, my research focused on the ancient Chinese monuments: I Ching (better known as the Book of Changes), Shan Hai Jing (or the Catalog of Mountains and Seas), and some others. As for I Ching, academician Vasiliy M. Alekseyev stated the following in his comments on the first translation of the Book of Changes by Julian K. Shchutsky: "Of course, you will not succeed in solving the mystery of I Ching, as it is insolvable until there is new archaeological data: after all, not a single text (in the full sense of the word) has yet been dug out! And, fragments are not text (this should be understood)." It has not been dug out to this today. Despite that research of the Book of Changes quite closely relates to the topic of Systemic Research in the Field of Human Psychophysiology, Alekseyev's comments prompted me to try researching other ancient Chinese books, approximately of the same dating, and to put aside the Book of Changes until a certain point.

The next subject of research was the ancient Chinese rarity—Catalog of Mountains and Seas (hereinafter Catalog). Inability to decrypt and use the Catalog led Confucius to that this catalog, due to its ancient origin, he classified as one of the sacred books of China and shielded it from research for many years, stating that a normal person should not communicate with devils and demons described in the Catalog. On the one hand, he closed the topic of research and on the other hand—Confucius issued a kind of writ of protection for the following centuries, which allowed to preserve and "...to bring to us the most ancient layer of mythological tradition, not darkened by subsequent buildups and re-comprehensions." (Electra M. Yanshina, *Formation and Development of the Ancient Chinese Mythology*.) Not only Confucius was powerless before riddles of mythology. Opinion of the famous ethnologist and scholar of religious studies James G. Frazer can be categorized as odd: "I hope that ... I will not be blamed for that I am a supporter of mythology, which I consider not only fallacious, but preposterous and absurd."

Returning to the topic, which Confucius authoritatively stamped as demonic, it is necessary to illustrate it first. "...The river of evaporations begins there, flows to the west and flows into the river Tan. There is an animal there similar to a mouse, but with head of a rabbit and body of a moose. His cry is similar to howling of a dog..." It is quite difficult to imagine such a monster. Or, for example: an animal, which resembles a rabbit, but has a bird's beak, eyes of an owl and tail of a snake. There are so-called "mythical" images, which are simply listed in Shan Hai Jing in a particular order and a specific scheme. Although it is necessary to note that until present time real, in our understanding, myths with action and storyline have survived in China, as well as in other nations of the world.

Research in the field of mythological images leads to interesting results, which sometimes, at first glance, seem absurd. Currently, it is acknowledged that the main peculiarity of a myth is that positions and information expressed in it are perceived by people as indisputable verities. These verities allow a person to have a holistic, noncontradictory picture of the world, in which a myth is characterized by remarkable stability and is reproduced with amazing precision, but requires eidetic (holistic) memory that is inherent in children and "wildman." By the way, even children of different cultural communities are very similar in terms of direction of their interests. As for us, adults, our passion for logico-discrete thinking leads us to recognition of the "...impasse of the so-called "magisterial way of development of world civilization," i.e. American and West European model of post-industrial society and the entire "spirit of capitalism," which stimulates newer and newer consumption and forms new and greater needs and attachments in the name of increasing production that brings new profits, again and again increasing the level of spiral of consumption.... This path cannot be imperative for all humanity..." because "...resources of the planet and its biosphere ... simply would not withstand..." ubiquitous standard of living of an average American and all of us "...would get horrors of Apocalypse." These are some excerpts from a wonderful work by Evgeny A. Torchinov titled Religions of the World: Experience of the Transcendence—Transpersonal States and Psychotechnics. However, for this report, the interesting moments in this book relate to research in the field of psychological approaches in religious studies and transpersonal levels of consciousness. Especially moments when transpersonal experience brings to the thought that was already voiced by Carl G. Jung: "Practically no attention was paid to that myths are first of all psychical phenomena, which express the deepest essence of the soul." At a certain point, this was that what prompted to consider the Catalog, conditionally called a mythological reference, as a system of archetypal images, which make up the "essence of the soul" of a human, but requires a holistic, eidetic perception.

It is necessary to remind of that chimeric constructs are used in the Catalog. All of them are original transformers, assembled from separate parts of animals and plants, on which research work was carried out to varying degrees and functional diapason of parts of chimeras is known. Description by our ancient ancestors of some particularities of character of a particular person followed by an attempt to holistically pass on information about these qualities to another person, who has never seen, for example, the person being described, is quite chimerical. Similar rudiments are observed today as well: dumb as a stump, a sly fox, a snake in the grass, strong as an oak, etc. More impressive chimerical examples can be found in various names of weaponry of all countries of the world. You know them perfectly well and it does not make sense to list them here. Even more refined actions related to use and artificial creation of chimeras are observed in advertising and cinema. A principled approach to the use of chimeras prompted to think about the possibility to consider functional psychophysiological diapason of a human through chimerical images. Another stimulus to such, seemingly paradoxical, idea were results published by Stanislav Grof and cases when people consciously reach states of consciousness, in which they self-identify with animals, plants, etc. These cases are well-known from the ancient shamanic, yogic practices, as well as from modern research. However, why identification by specific people was with some specific animals and plants? Why did other subjects identify with other animals and plants, often with chimeras? There must be a certain scheme in this. First of all, I turned to phenology. After turning to the phenological studies, it could be summarized that "even the stork in the sky knows its seasons, and the dove, the swallow, and the crane observe the time for migration. But my people do not know the requirements of the Lord" (Jeremiah 8:7). "Fresher" research, for example, migrations of animals, showed that "physiological stimuli, which cause migration and reproduction, are determined by the interaction of biological clock with annual cycle (biological calendar) and phytoperiod measured by biological clock with 24-hour cycle" (John Cloudsley-Thompson, *Animal Migration*). I hypothesized that "inset" of functional diapason of psychophysiological structure of a human is somehow connected with phenological algorithms. Regardless of whether he wants to or not, a human obeys the existing laws of vital functions of biogeocenosis. Research in this area illustrates this quite impressively. For example, reproductive processes of various species of fauna and flora obey the phenological schedule. In any specific period a certain form of species is born with a specific functional diapason, by researching which we have classification systems of the organic world. In the case of a human, we have basically one and the same form, but different patterns in characteristics of the functional diapason of the form, which, by all appearances, was determined in the ancient times by chimerical formations, such as in the Catalog of Mountains and Seas. It was necessary to identify the algorithm of distribution of the human species and description in the Catalog in phenological cycles and then—compare. The development of this hypothesis led to the distribution of chimeric complexes of the Catalog according to the annual cycle. The same was done with a human.

There was the final stage left—experiment. Experiments were carried out. Jumping ahead, it is possible to say that results of experiments were somewhat striking.

In them we began to receive data, which had to be distributed somehow. Six factors were introduced for this: intellectual, physical, nutritional, sexual, emotional, and environmental. The intellectual factor accumulated information that characterizes the power of intellect, its ability to accumulate and process information, generate ideas, perform analytical work, and so on. The physical factor included information related to the attitude of the research subject to his physical parameters, hygiene, field of applications, texture and color characteristics of clothing, etc. The nutritional factor summarized information related to preferences of a specific menu, frequency and form of consumption, and so on. Information in the sexual factor included data on power of potency, frequency and ways of sexual relations, choice of monogamy and polygamy, sexual orientation, attitude towards offspring, and so on. The emotional factor contained information about the emotional algorithms of the subject of research. Finally, the environmental factor considered facts of education, profession and career choice, and well as choice of interior, migration and other processes. Naturally, these factors were considered in time interaction: daily, monthly, seasonal, annual cycles.

Experiments included audiovisual division between research subjects and the analytical group, which used the Catalog, and those, who communicated with research subjects independently and worked on the basis of existing psychological methods. Results of experiments made it possible to draw conclusions. First of all, information obtained by testing covered psychophysiology of the subject of research at the same time instant and not in any one of phenological cycles. Secondly, there was no classification grid in distribution of information received by psychologists. Thirdly, there was significantly more information obtained using the Catalog compared to information obtained by the method of testing and there was data that did not relate to psychology. Fourthly, as it turned out, research subjects themselves belonged to different groups: 1—those who never wondered about patterns of their functional diapason; 2—those who wondered about this question when necessary; 3—those who considered this question completely unsolvable in principle. Proceeding from this, sources of identification of received information using the Catalog could be permanently realizable physiological desires of research subjects, which they themselves highlighted as vital and top-priority, as well as data from the group that communicated with them with the goal of testing. It was the comparative analysis of information obtained from experiments that was somewhat striking for me. Vital psychophysiological needs of test subjects completely matched data obtained using the Catalog, were supplemented by this data about seventy percent and were perceived as a guide to desired actions. To some extent there was a match with materials of testing, although there was significantly less volume of information obtained in the latter and they were formulated somewhat amorphously, which in general is characteristic of psychological products of similar kind.

Now, moving on to manipulation modes; by manipulation I mean natural and artificially creates schemes, application of which puts the research subject in obedient, equal, and leading positions and when combined causes actions that research subjects themselves consider necessary, but which are already prognosticated and described in advance in a certain hierarchical scheme.

Allow me to tell a little about where the idea of manipulation modes itself came from, or, to try to put it another way, about interaction of different systems in a manipulation mode. For me, it was of great interest to try to overcome some discrepancies between information received with the help of the Catalog and other methods, which, as it seems to me, makes sense to balance both in quantitative and qualitative aspects. Manipulation in the conceptual framework that I allowed myself to define could become that tool, which would provide an opportunity to fill the manipulation scheme as a whole and provide additional information. What I mean here is not selecting an individual for a so-called clean experiment, but rather considering a human and the environment (system) surrounding a human, including social, as a single system to begin with. And, this is what was done.

As it seems to be, now it would be more appropriate to discuss subtype characteristics of human species because obtained information lies in psychophysiology of a human as a whole and not in psyche and physiology separately. Based on obtained information, it is possible to discuss characteristic patterns of functional diapason of human form in different time cycles. This gives reason to consider that it is necessary to divide human species into certain groups of people—carriers of characteristic patterns. Understandably, it is logical to call such groups subtype groups because the term "subtype" is a taxonomic unit, a rank lower in the taxonomy of plants and animals, but represents totality of geographically separated populations of the same type, representatives of which differ from representatives of other populations of the same type by specific, rather stable characteristics—this is the goal that was pursued in the title of this report.

In conclusion, I would like to say the following: if part of the organic world, to which abovementioned animals and plants belong can be considered bio-forms with software, by which classification as such is made, then a human can also be considered as a programmed by nature part of this world.

I think that possibilities that open us with the help of this research could to some degree influence the state of affairs of Russian Federation. If basing on that the main resource of civilization is a human, his abilities and characteristics, then a more adequate use of this resource can provide a more goal-oriented vector of the development of civilization. Especially since in interaction of characteristics of being studied patterns of such a scheme as Universe—Earth—Human, a Human becomes a better known component.

References

Davydov, A. (2002). *Katalog Chelovecheskoy Populyatsii* [*Catalog of Human Population*]. Presentation at the International Conference on Prospects of Preservation and Development of Unified Civilization of the Planet: Culture, Ecology, Cosmos (in Russian), Moscow (2002). [Mention in *Parlamentskaya Gazeta*: "...The most luminous were reports by ... A. Davydov about the program that he developed, which opens up the best, but often hidden qualities of a human..." (Umorov, I. (2002). *Chelovechestvu Otpushcheno Tol'ko Tri Goda* [Humanity Has Only Three Years]. Moscow: *Parlamentskaya Gazeta, Sphere of Reason*, 129 (1008), para. 4. Retrieved 13 May 2015 from http://old.pnp.ru/archive/10082921.html.)]

PART 2

SCIENTIFIC ARTICLES

ARTICLE 1

FROM CARL GUSTAV JUNG'S ARCHETYPES OF THE COLLECTIVE UNCONSCIOUS TO INDIVIDUAL ARCHETYPAL PATTERNS

Andrey Davydov and Olga Skorbatyuk

Translated by Anonymous.

Edited by Kate Bazilevsky.

It is a well-known fact that neither Sigmund Freud nor his student Carl Gustav Jung were the first scientists to start their quest in the field of the unconscious. Prior to them Carl Gustav Carus and Eduard Von Hartmann had worked on the philosophy of the unconscious, though the very idea, was once ousted by the fashionable empiricism and materialism of the time – the concept being naturally scientific in psychology. However, despite the greatest discoveries in this area, the soul, as the basis for human psyche, was not only failed to be found, but its very existence was denied. The results of this fact might be well observed nowadays in psychology, philosophy and medicine: no soul – no foundation, no foundation – no house, no pivot – nothing to put the rest on. And, possibly not knowing the modern human being is the very cause of massive fragmentation or splitting of the conscience regardless of the diagnosis.

True we could be quite content with what we have already discovered in the fields of psychology, philosophy and religion, but these findings don't make us fully satisfied. Also, they don't explain who we are, why we are in this world and what our purpose is. In other words, they don't answer the questions of vital importance. In fact, we don't expect to be truly happy with answers to the most important every day questions. And, the proof to this lies in general disharmony in various spheres of our lives as well as the very fact of continuous search for human soul.

We were quite positive that it was the profound or "deep" psychology that could help us in this search, the reason being the very fact that whatever "lay on the surface" had already been discovered, though unfortunately was not worth discovering. In Jung's view deep psychology was similar to archeology and prior to him Sigmund Freud many a time compared psychoanalysis with this science as well since according to him Jung dealt with the "excavation of soul". If "archi" means "origin," then deep psychology should deal with "archeological excavation" of human psyche rather than rocks, penetrating the former as the basis of consciousness. After all, psychology like all other sciences is meant to study the universal in the particular, trying to find common patterns in something individual. However, it is well-known that common patterns do not lie on the surface therefore scientists try to get to the bottom going deeper and deeper.

So, we've decided one more time to try and find the very basis of human psyche, the cornerstone, which gives rise to the rest of the structure. And, until the end of the tangle is found we won't be able to unwind the latter in order to understand what human psyche (which someone may refer to as soul) is all about.

Since the main scientific source for our psychological research are materials primarily borrowed from Carl Gustav Jung (1875-1961), we would like to focus our attention on the works of this great Swiss psychiatrist, psychoanalyst and psychologist. It is important for analyzing his conception of archetypes and their effect on human psyche. Examining his theory, our task will be to determine in what part our views on the subconscious are the same as those of Jung's and where they are cardinally different.

Before we start discussing archetypes both in general and from different angles, it would be only logical to give them proper definition. The noun "archetype" comes from two Greek words: arche – origin and typos – form, pattern. During the later period in ancient philosophy the above concept was used to describe the prototype or idea; this was primarily an original model, a newly created pattern. One of Jung's works contains the following: "...the term "archetype" occurs as early as Philo Judaeus, with reference to the Imago Dei (God-image) in man.

It can also be found in Irenaeus, who says: «Mundi fabricator non semetipso fecit haec, sed de aliens archetypes transtulit» [(lat.) "The creator of the world did not fashion these things directly from himself but copied them from archetypes outside himself."] In the Corpus Hermeticum, God is called το ἀρχέτυπον φώ ς.. (archetypal light). Although the term "archetype" is not found in St. Augustine, the "idea" of it is. According to Jung the word "archetype" is an explanatory paraphrase of the Platonic εἶδος (eidoV). For our purposes this term is helpful because it tells us that we are dealing with archaic or rather – primordial types, i.e. with universal images, which have existed from time immemorial (*Archetype And Symbol*, 1991, p. 98). The term "archetype" is similarly used by alchemists and Jung himself refered to Vignerus who states that the world is «ad archetypi sui similitudinem factus» [(lat.) "Made after the likeness of its archetype" Tract. De igne et sale // Theatr. Chem., 1661, VI.3]

According to Carl Gustav Jung the original ancient psyche lay foundation for our intellect in the same way as our body structure goes back to general anatomy of mammals. However, when discussing this foundation he meant unconscious ideas and images, which are part of the collective unconscious. To Jung the same archetype could be equally applied to whole nations and eras.

After conducting research on Greek mythology and analyzing the dreams of mentally ill Africans, Jung came to a theory that major mythological motifs are common for all races at all times. He thought of the collective (or "racial") unconscious as hereditary, interpersonal and resulting from human evolution. The contents of the former are called archetypes. By Jung's own definition an archetype is "a symbolic formula functioning whenever the concept of the conscious is either yet non-existent or altogether impossible," it's a certain "instinctive vector, a pointed trend, the one, which forces birds to build their nests and ants to erect anthills" (*Archetype And Symbol*, 1991, p. 65). We consider the above definition to be quite accurate, although, as opposed to ants and birds, the author of analytical psychology for some reason denies humans their own "individual destination, with no connection to mankind's general purpose as a species. While agreeing with the existence of Jung's archetypical matrix as "infinitely ancient in psychic origin," we still feel that it contains images of an individual archetypal human pattern, rather than those of the collective unconscious, common to everyone.

Jung believed that people are born with most hereditary traits, inherited from their ancestors, these traits being a main factor in controlling people's behavior and partially in determining their character, psychic structure and possible reactions. We, on the other hand, think that the collective unconscious, though not without its obvious effect, does not control one's behavior and does not determine one's reactions and psychophysiology on the subconscious level. In our opinion the main factor in structuring human's subconscious psyche and determining his identity (behavioral pattern or cognitive strategies) are not the values inherited from our forefathers and therefore not the ones capable of serving as key elements in forming his personality. After all general ideas and images by definition can't create individuality – that would be contrary to logic. Jung discussed the existence of some racially-preformed collective persona, which selectively penetrates into man's world, can be modified and further developed. Somewhat supporting Jung's point of view on this issue, we still must clarify the matter in that the extent of penetration of this so-called "collective persona," which is based on archetypes of the collective unconscious, in our opinion can be controlled and held in check depending on each individual's archetypal pattern. Since if we were to suppose that the hereditary unconscious ideas and images happen to be common to everyone, then all people would demonstrate exactly the same "standardized" reactions to those images. However, is this what we see in reality? True, more research is needed here, although from what we have observed, it is quite obvious that people react differently either to the image of Jungian Mother or to that of the Trinity. We feel that, despite the importance of racial (collective) archetypes and their undoubted influence, the degree of this influence on each person would not be the same. It depends on his personality, which is determined by a number of his individual archetypical images, and just like with a black box, information going through an individual on the input will be different on the output. Otherwise, a human and his spiritual life would be totally inseparable from the society. We certainly can't deny frequent occurrences when a society affects people and their lives, but this happens primarily due to their lack of knowledge of themselves and their personal archetypal patterns, not because of a strong influence of the archetypes of the collective unconscious.

So, where all this information concerning a person's individuality could be stored? Let's make an attempt to find it in Jung's works. Human psyche or, as Jung has put it, his soul consists of three differentiated but interconnected systems. With his conventional approach of presenting human psyche in two spheres – the conscious and the unconscious one, Jung divided the latter into a system of the personal unconscious and the archetypes of the collective unconscious – these two segments being completely different from one another. "Ego" as conscious mind, made of supraliminal perceptions and ideas and responsible for human's identity is considered the center of the consciousness. We will not analyze the conscious sphere since enough attention has been given to this issue by other psychologists. "What we call "psyche" or "soul" is in no way identical to what is contained in our conscious mind" (*Archetype And Symbol*, 1991, p. 27), where intellect, conscious ideas and attitudes, like many other notions, in no way associated with soul's irrationality, are located. Jung emphasized that the wholeness of the soul cannot be detected just by intellect. "Life is logical and illogical, rational and irrational... Psychology that caters to intellect alone is not practical..." (Jung C.G. Über die Psychologie des Unbewuβten). So, "ego" seems to us as a fairly clear system analyzed enough and being in no need of any additional study for the time being, while the collective unconscious and its archetypes are universal and stand completely apart from the personal in the life of an individual. So, where is the soul? Now it's necessary to focus on an issue as to what Jung really meant by the "personal unconscious" as part of the soul structure.

Jung defined this segment of human psyche as a certain region adjacent to the Ego. He feels that the personal unconscious contains ideas and emotions which at one time used to be conscious, but at present are repressed, forgotten or for some reason purposefully ignored or forced out. It's a certain dark corner of our subconscious "Ego" (which Jung himself called the Shadow) where all forgotten and repressed contents "live the lives of their own," and which unlike the collective unconscious, consists of one's personal experience, acquired during one's life time. Naturally in the light of our research and new findings we beg to differ with this assumption.

According to Jung, the realm of the collective unconscious also includes people's complexes. We perceive the personal unconscious complex as a combination or constellation of feelings, ideas, perceptions, memories, etc. This complex contains a nucleus, which like a magnet attracts various emotions, as if "constellating" the latter. If we take the "mother complex" whose nucleus is created due to mother connected both racial and child emotions, then all thoughts, feelings and memories related to mother will be attracted to this nucleus, creating a complex. And, the more powerful the force of the nucleus the more emotions it withdraws. According to Jung, a person, dominated by his mother, will possess a strong mother-complex. And, in this case, all his thoughts, feelings, actions, etc. will be led by mother filled conceptions, her image being superior in his personal unconscious. Jung warned that an image-forming complex can not only effect a personality, but can easily capture it by using psyche for its own purpose, as with Leo Tolstoy and his forgiveness-dominating idea or Napoleon, obsessed with his thirst for power. And here, everything seems to make every bit of sense: the more dominating the image, the more it controls a personality, somewhat molding the latter. On the other hand, it turns out that the images of the personal unconscious merge into the collective archetypes (as with the Mother-merging image), and here again Jung put forward racial or collective emotional experience as the main complex-forming factor. Then, what makes the personal and the collective unconscious different from each other if they are both created by images originating from the same source, same archetype. Could this mean that these two spheres are not quite differentiated, which results in their partial merging? If this is true, why then would the author state that the collective unconscious is universal and separated from the personal in the life of an individual? Even Jung himself defined the instances, where the archetypes of the collective unconscious "overlap" the ego and the personal unconscious as pathological right from the start. Can we then consider as normal the fact of the collective unconscious "mother-image" overlapping the area of the personal unconscious or it would be pathological? Taken as true, this assumption should probably mean that pathology is initially embedded in the personal unconscious by the very possibility of merging with the images of the collective unconscious? However, it makes no sense.

The way we see this whole mess is as resulting from misunderstanding the contents of the personal unconscious and questioning the latter's very existence and what actually constitutes a personality. Could the essence of the latter be based on various thoughts, feelings, memories, a child's emotions, etc.? And, could in fact such short-term psychological phenomena lay the foundation for a stable system? After all any personality always presents a certain system, something constant, the core, though being somewhat modified due to one's life situation, but not with a speed of continually changing thoughts, feelings and emotions.

Jung clarified the above statement in that the collective unconscious is universal due to the fact that all races show similar brain structure resulting from evolution's generality. Fairly logical as it may seem at first blush, this doesn't explain why, despite the collective archetypes' effect on an individual on a subconscious level and that of social stereotypes' on a conscious one, he demonstrates such a great variety of reactions to collective values. Jung maintained that racial memories or representations are not hereditary as such: more likely we inherit the very possibility of reliving the experience of previous generations and they function like inclinations which make an individual react to the world in a manner as if inherited. Then, why some individuals are willing to take advantage of their forefathers' experience, while others are not? What factor determines if a human should make use of his "heritage" or act in his own original way? Here, we are not considering his instincts or reflexes, which might be quite appropriately added as part of the collective unconscious being common to all types.

Since humans to some extent must carry a burden of racial feelings and collective archetypes as well as the complexes of the personal unconscious, will they still be regarded as personalities or at least sound individuals without the "pathological overlapping." As Jung stated the collective unconscious is (no more no less) an inborn racial basis for personality's structure. And, according to Jung it cultivates the personal unconscious and other individual acquisitions. Well, it seems that even what an individual believes to be the result of his empirical experience, is actually determined by the collective unconscious and its chief or selective influence on the individual's behavior all through his lifetime.

However, is that really the case? It's hard to dispute the fact that since all human beings have always had a mother, every infant is born with certain instinctive perceptions of her. However, it's not just characteristic of human species. A child who is regularly given milk from a bottle with a nipple by an orphanage worker may confuse the latter with his mother just like a famous gosling (classical example from animal psychology) taking every toy or a ball for his mother and following it everywhere. Jung asserted that this acquired awareness of one's mother, so crucial for an individual, could be the realization of his inherent potential, "built" into human brain by his past racial experience. Possibly it is so. However, how does this make us different from animals, the latter having no sign of the collective unconscious, but simply possessing inborn instincts and reflexes just like humans.

If we were to accept Jung's theory without reservation, our perception of the world is primarily formed by the collective unconscious, though, according to his own stipulation, only partially; otherwise either no variations for development or the development as such would be possible. Assuming that an individual's right from his birth is assigned such a modest role in the general order of collective values and influences, where is the room for his personal development? However, this would be somewhat contrary to the logic of nature, which, metaphorically speaking, either "waits for" and sometimes "thirsts for" or even "demands" personal development as the key link in the chain of humanity's general evolution, and without which our life on planet Earth would lose its meaning. Following this logic, nature should have provided enough opportunities for human development, growth and personal evolution. However, here the individual space is narrowed to its maximum with the contents, introduced by someone from the previous life, i.e. the contents of other people's past experience. So, we are dealing with a conflict: on the one hand passing on former generation's experience is important in that it saves strength, time and in some cases even descendants' lives. Not to mention the role of generation's continuity in passing on cultural values, culture being indispensable for personal development. On the other hand, it reduces individual space that serves as a foundation for building a personality.

After finally examining the psychic structure at Jungian school, we still failed to find either the room where individuality should be stored or the material, which the latter is "built" of. We couldn't answer the question as to the "location" of the Jungian unconscious and the form in which the individual's personal qualities are recorded, the latter being neither hereditary manifestations or semi-pathological complexes, nor passing thoughts. True, the unconscious possess the centuries-old wisdom and experience of past generations, but we believe that the collective archetypes, nucleus-containing complexes, childhood memories and thoughts interspersed with feelings, can't serve the determining factor as to the structure of human psyche or the contents of this structure. Not only this would make sculpting" a personality impossible, but create a problem in just singling a unique individual out of the dull crowd of human specimen, i.e. to make a clear distinction between people. Another thing would be to see the correlation between hereditary genetic perceptional and behavioral patterns and the ones inherited due to cultural and historic memory. In their attempt to clarify this problem psychologists might expect to be assisted by experts in such fields as: anthropology, ethnology, linguistics, ethology and history.

Thus the collective unconscious as the focus of Jung's works is where the archetypes are stored. According to Jung, these archetypes sort of sum up the life of a species, so the collective unconscious being an integral part of any person's psyche is hereditary and serves as a foundation for growing an individual psyche. We totally agree with Jung's theory that the basis of psyche is located in its unconscious sphere and don't dispute the fact that human mind during the evolution accumulated some specific reactions common to everybody and established as instincts. However, we feel that the cornerstone of human psyche is not the collective unconscious, but individual archetypal patterns. To us they are contained in the personal unconscious, if to use the Jungian terminology, and as logic has it, the collective unconscious can contain just instinctive programs along with some cultural experience, which should belong to a human as a species for his personal needs, understanding, etc.

No doubt archetypes form the basis for human psyche, and if we perceive them as some unconsciously reproduced stereotyped patterns presented as myths, fairy tales, hallucinations and works of art (which is contrary to general opinion), there would be nothing unusual or mysterious about it. Though as Jung has put it: "...while interpreting the myths anything but the soul was appealed to..." (*Archetype And Symbol*, 1991, p. 10) With such an approach in mind it is clear that human psyche is not "tabula rasa" (clean board) on which anything can be written down just by anyone..." In trying to identify the psyche and the consciousness one can be easily driven into the false idea that a person enters this world with empty psyche and later in life his psychic will contain nothing but what has been attained with his personal experience" (*Archetype And Symbol*, 1991, p. 98).

Jung maintained that patterns of the collective unconscious are innate elements of the human mind, much the same as instincts; acquired at birth, they begin to manifest themselves in much the same way in all individuals. This is defensible only in light of the author's assertion that collective patterns are attributes of the human mind. The mind, however, should refer not to the unconscious, but rather to the conscious realm wherein personality is decidedly influenced by collective thought.

So, what does the unconscious have to do with it? In Jung's opinion "the inner motives arise from some deep source with no connection to the consciousness and not controlled by the latter" (*Archetype And Symbol*, 1991, p. 76). Due to the cultural revolution human mind made quite some progress in getting to know the outside world. However, it didn't come one iota closer either to acquiring the knowledge of itself as inner substance or the knowledge of the foundation that holds and initiates its own psyche. At one time Jung characterized this situation in the following way: "Our intellect has achieved the most tremendous things, but in the meantime our spiritual dwelling has fallen into disrepair. We are absolutely convinced that even with the aid of the latest and largest reflecting telescope, now being built in America, men will discover behind the farthest nebulae no fiery empyrean; and we know that our eyes will wander despairingly through the dead emptiness of interstellar space" (*Archetype And Symbol*, 1991, p. 107). And, further he added: "While becoming more scientific our world becomes more dehumanized" (*Archetype And Symbol*, 1991, p. 86). These words are the evidence of his clear understanding the problem. However, his main accomplishment was in that he didn't just see the problem, but being intelligent and well educated he knew the way to solve it: "We are finally turning to the wisdom of all times and all nations only to discover that the most valuable words were actually delivered long ago in the finest language – the language of symbols" (*Archetype And Symbol*, 1991, p. 107).

In our opinion, Jung's mistake was in that information recorded in the unconscious isn't just of the universal, but also of individual origin, as with instincts and other cultural traditions. And, it's probably in this part where lies the main distinction between our perception of the basis for the human psyche and that of the author's of analytical theory, who was first to bring up the concept of archetypes in psychology. This is what he says: "...I have chosen the term 'collective' since this part of the unconscious is not individual but universal. And, unlike the personal soul its contents and modes of behavior are relatively the same everywhere and in all individuals. It is, in other words, identical in all men and thus constitutes a common psychic substrate of a super personal nature, which is present in every one of us. We will make an attempt though to oppose Jung on this issue. He stated that "similar programs and common patterns affect not just some primitive modes of behavior like unconditioned reflexes, but also our perception, thoughts, imagination" (*Archetype And Symbol*, 1991, p. 14). Here is the difference: Jung considered the archetypes of the collective unconscious to be some cognitive patterns the instincts being their correlates. We, on the other hand, define these archetypal patterns as those of the individual nature. They affect human's behavior and psychophysiology while instincts and reflexes are universal and hereditary. Jung was right in that "grasping the archetype comes before the act, triggering the inner behaviour," but if the archetypes were universal the same stimulus would make them cause identical reactions in people (*Archetype And Symbol*, 1991, p. 14). And, that would be contrary to both the common sense and reality.

We accept Jung's assertion that the archetypical matrix, the one that a priori shapes human fantasies accounts for common motives in myths and fairy tales, explains the appearance of made-up characters. The large question is which archetypes and images underlie myths? "Primitive man impresses us so strongly with his subjectivity that we should really have guessed long ago that myths refer to something psychic" (*Archetype And Symbol*, 1991, p. 100). Assuming that mythology actually is of psychic origin, it would be logical to ascribe it to one's individual psyche, i.e. to the archetypes of an individual nature. The fact of the matter is that "myths are first and foremost psychic phenomena that reveal the nature of the soul" (*Archetype And Symbol*, 1991, p. 99). In Jung's view, the archetypical images have been throughout the centuries the main material for creating central figures in religion, mythology, works of art, etc. However, the gruesome and disarrayed images of the unconscious, portrayed as the contents of the creative mind, are subject to gradual refinement and transformation into symbols beautiful on the outside and universal on the inside. Jung defined mythology as "the tool for initial shaping the archetypical images" with an archetype at the very core to be first polished beyond recognition. Curiously the finer the image, the farther it's distanced from one's personal experience and hence less recognizable to others, overpowered by endless comments of the crowd. This is the mechanism as to how the personal unconscious of an individual nature is transformed to the universal mythological images.

With time passing by the archetype has grown less archaic which was evident through ancient drawings where primitive people portrayed animals not so much like symbols, but rather as careful and detailed pictures. In Jung's view, when at the dawn of our civilization the human mind was mainly ruled by instincts, man's consciousness might have grouped them into some models or patterns. Ours is a belief that prehistoric man was very close to nature and of as yet "uncorrupt mind". What really matters is not the fact that he arranged his instincts in patterns, which supports Jung's theory, but he was instinctively capable of "reaching" these patterns in his personal unconscious. Creating the drawings not of the symbols, but of actual animals, he viewed them as god-like objects of worship and learning, with full and clear understanding that he is the one most defenseless of the living creatures, sharing psychological ties with all of them. In our view, it is one's individual psyche that has the closest connection with nature and its components (animals, plants, minerals and natural resources) containing the records of the latter through the archetypes as their original patterns. Hence, the thought: in trying to explain the symbols and their archetypes one should do so from the stand point of nature and only then applying Jungian approach of socializing some concrete images.

Note that the human situation actually weakened due to the growth in technology and its "confusing influence" on archetypes. Jung indicated that the life of modern man being overpowered by the cult of rational thinking isolates the conscious from the unconscious, leading to the natural imbalance between rational and irrational processes. As a result the unconscious "invades" the realm of the conscious mind to compensate for latter's limitation. As Jung has put it: "The conscious mind ignoring the experience of the archetypes may result in their invasion its territory in the most primitive forms causing wars and social upheaval" (*Archetype And Symbol*, 1991, p. 17). "Modern man fails to see himself as victim of his own 'rational attitude' thrown at a mercy of psychic inferno" (*Archetype And Symbol*, 1991, p. 85). By "psychic inferno" Jung probably meant man's overall confusion and isolation from our mad world with its wars, revolutions and mass psychosis. If Jung attributed the cause for social disasters to the "intrusions" of the collective unconscious, ours on the other hand is a firm belief that the latter in no way could result in the appearance of collective myths of any sort. We feel that mythology was mainly the product of some historical characters' creative mind with their own individual archetypes, who through their ideas passed them on to congregations causing violence and collective madness amongst them.

This has been happening due to man's inability to see how symbols affect his conscious mind lacking both the knowledge and the inquisitiveness as to the nature of these symbols. Note that man normally doesn't regard symbols as an object for analysis: he prefers to rely on their interpretation by social institutes whose sole purpose is to influence his mind to their advantage. Living in terms of symbols man unfortunately neglects to look into their origin. Considering the fact that Christian symbolism and customs having had their impact on the human mind for about two thousand years, it would seem odd to see the majority still refuse to see into their meaning. While describing the religious symbols, Jung stated that people accept them with no doubts or reflections much in the same way as they view the process of decorating a Christmas tree or coloring Easter eggs not having any idea as to its meaning: "It would be futile to ask a modern man about the true meaning of a Christmas tree or Easter egg. The fact is that people do certain things without knowing the reason for their actions" (*Archetype And Symbol*, 1991, p. 7). Though being not quite inquisitive about things could be dangerous for a person since he may get involved into unclear introjections of suggested symbols. However, according to Jung, what's important is that he who realizes his "spiritual poverty and the loss of symbols," but is incapable of being satisfied with just the 'ersatz', may find himself in rather a complex situation: "frightened he turns away from the nothingness that shines in front of him. To make the matter worse the vacuum is filled with political and social ideas marked by absurd and spiritual emptiness." People not knowing who they are and therefore lacking their own identity become part of a crowd that is easily controlled and follows any idea. And, this is exactly what's been happening: people like a herd follow the "Sheppard" wherever he leads them. And, the question is only in WHO is the "originator" of the idea. That would determine the crowd's direction: whether it be it the "bright communist future" or the gas chambers of fascism. One way or the other, the human is always controlled by the collective unconscious, myths and symbols created by someone in the past.

Mythology was first used to help primitive man to explain natural phenomena, but more so it helped him live through the discrepancy between his "evolutionized" conscious mind and that of the unconscious with its animal-like horrors and fears. Come to think of it this very gap between the conscious and unconscious spheres is presented in the Bible as the Fall and the Expulsion from Paradise, and in his attempts to bring back the lost harmony human turned to mythology, fairy-tales, magic, etc. Ironically, the evolutionary process only increased the chasm between both spheres, demanding more myths not just for dealing with his intense fear and anxiety, but with the continuous sense of insecurity, the issue being—adjustment to his inner world. This gave rise to modern mythology with stories not of heroes, but the cellulite since the latter is more vital to modern man whose inside is filled with gaping emptiness. Science and technology being the product of the conscious mind may be regarded as a tool to keep reality in check and whose function is akin to that of creating mythology: "Efforts that modern man exerts in science and technology were in prehistoric times applied in making up myths" (*Archetype And Symbol*, 1991, p. 17). No technology though is capable of providing him with by far the most comfortable feeling: that of inner stability, harmony and peace – features characteristic of an integrated personality well in tune with both its conscious and unconscious mind.

Disillusioned by all castes of experts on psyche to find his personal soul, modern man turned away from his inner being to fully concentrate his efforts on getting to know the outside world. This resulted in his transformation into a confused, sick and miserable being with anxieties, fears and panic attacks, uncertain about his belief in that the thinking mind should be of unlimited capacity and superior to all. The realization that "all-powerful human mind" has turned out to be another myth, was a frightening thing to take: a human always faces things beyond his control be it the natural phenomena or his own archetypes. Jung confirmed that "whether primitive or not, mankind always stands on the brink of actions it performs itself but does not control" (*Archetype And Symbol*, 1991, p. 113). Intellect put a human up above nature only to throw him down off the pedestal later, making him realize that he was no longer in control of his own soul, his mood and emotions, or capable of detecting any loopholes which his unconscious could leak through and influence his actions and behavior.

Throughout the course of history man along with social institutions concentrated his efforts on perfecting his conscious mind. The same goal was attained by means of religious dogmas and worldly rituals serving as walls to protect human psyche from "direct collision with archetypes' colossal psychic energy" (*Archetype And Symbol*, 1991, p. 18). The visual effect was that of the struggle with obsession, "release from spell," exorcism, redemption and purification. Jung stated that "man has never had a shortage of powerful images, which like a magic wall, protected him from the eerie vitality hidden inside his soul" (*Archetype And Symbol*, 1991, p. 104). Jung illustrated it by a story about a Swiss mystic and hermit who experienced the so-called Trinity vision while in his solitary cell at Sachseln. As the story goes brother Nicholas "gazed deep into the dark mirror, which reflected the wondrous and horrifying light of the primordial" (*Archetype And Symbol*, 1991, pp. 101-104). Besieged by his own unconscious, he must have turned to the dogmatic image of divinity "for help in assimilating the fatal invasion of the archetypical image" (*Archetype And Symbol*, 1991, pp. 101-104).

Jung maintained that "dogma substitutes for the collective unconscious by formulating its substance in man's thinking mind. The life of the collective unconscious has been presented by archetypical dogmatic ideas and continuously flows amongst the Credo symbols and ritual" (*Archetype And Symbol*, 1991, p. 101-104). Ours is a comment that it makes every bit of sense for people to sacrifice their identity along with possible personal growth and perfection for just the illusion of "peace and security" rendered by religion or its like. On the other hand, the fee might be way too high. Thus, the "Faust's" notorious Mephistopheles, demanding selling the soul in return for certain comforts, is nothing unusual since all gods, be they Christian or not, act in exactly the same way. As a result, the individual archetype is hidden by the collective and man goes on "living in his cave gazing at the shadows on the wall". [The reference is to the Platonic myth where people were held captive in the dark cave unable to escape the womb of the Earth trying to reach for the Sun. (Republic, VII, 514-517)] However, why do "Gods die" if man is still capable of using the archetypes of the collective unconscious like The Great Mother or Trinity as a shield against the images of his personal unconscious so terrifying, uncontrollable, and unpredictable. In one of his works Jung made a comment that "the gods die from time to time due to man's sudden realization that they don't mean anything, they are made by his hand of wood and stone and are totally useless" (*Archetype And Symbol*, 1991, p. 105).

Someone's ideas may attract man so as to temporarily fill the emptiness inside him. Besides, their introjection is fairly simple and does not require any special creative efforts: no need to get into heavy thinking, take risks, or go through any painful experience. The values of the collective unconscious are like cooked dumplings – they are easy to use. However, they give rise to new patterns pertaining to anything from primitive emotional manifestations to one's views and outlook. Due to his stereotypical behavior, actions and thoughts modern man can save a lot of time to accomplish a great number of things not necessarily related to him personally. However, irrespective of that there still exist some steady behavioral patterns as to the way he should live, work, feel, etc. not willing and having no time to spare both his mental and physical efforts.

There's a standard list of social recreational activities in which one gets involved during one's personal free time, and the choice extends from embroidery to parachute jumping and from attending theatre to training seminars on Tantric sex. Modern consumers readily consume both material and spiritual values generously offered by concerned parties – the fact well described by psychoanalysts and philosophers in their works. This is how our social life is organized and understood and in this sense either highly developed capitalism or pluralistic democracy with its variety of "existence" (not life) opportunities are not much different from totalitarian egalitarianism. By far not everyone is willing to spend his time and efforts on self-improvement and people for the most part are quite content with whatever values the society has to offer. However, as the old saying goes: "a sacred place cannot stay unoccupied for long" and if a person doesn't feel like improving his inner world to know himself better, someone else will gladly do it for him. After all, human material can always be used for any purpose.

However, this is by far not the only danger. Besides someone else's will, there's another very important factor affecting us, and which we know nothing about–our own unconscious. As Jung has put it: "In our mind we are our own gods, the 'factors'" [here the author refers to his statement as to gods being called "factors" from facere – to make.] (*Archetype And Symbol*, 1991, p. 113). "But once we have walked through the door of the Shadow, we discover with terror that we are the objects of unseen 'factors'. It is really unpleasant to know this since there's nothing more disillusioning than finding out the truth of our own inadequacy" (*Archetype And Symbol*, 1991, p. 113). Jung further maintained: "Ignorance is no guarantee of security, and in fact only makes our insecurity still worse, it is probably better despite our fear to know where the danger lies. The greatest danger threatening us comes from the unpredictability of the psyche's reactions" (*Archetype And Symbol*, 1991, p. 113). Every adolescent clearly sees the way out of this "spiritual deadlock": a person can't know the world without knowing himself first, which means that he has to admit his "poverty" and take the situation under control. No matter how hard this admission might be, man nevertheless keeps striving for self-knowledge making useless attempts to "find" himself through arts, religion, psychology and mystics. Everywhere he faces common patterns so different and yet so similar.

Besides, in search of himself man confronts another big danger. Jung stated that "whoever looks into the mirror of the water will first of all see his own face. Whoever goes to himself risks meeting with himself (*Archetype And Symbol*, 1991, p. 113). True there's always a doubt as to whether fear is the only obstacle for man on his way of researching deep into the "dark waters of his unconscious". Perhaps, not fear but pride is one of the worst sins! "...The mirror does not flatter, it faithfully shows whatever looks into it; namely, the face we never show to the world because we cover it with the

Persona, the mask of the actor. However, the mirror lies behind the mask and shows the true face. This is the first test of courage on the inner way, a test sufficient to frighten off most people, for the meeting with ourselves belongs to the more unpleasant things that can be avoided so long as we can project everything negative into the environment" (*Archetype And Symbol*, 1991, p. 113). It would seem like strong-minded natures are capable of almost anything, but with respect to "meeting" with their own unconscious their strength may work against them: "The Shadow reminds man of his helplessness and ineffectuality. Strong natures – or should one rather call them weak? – do not like to be reminded of this, but prefer to think of themselves as heroes who are beyond good and evil, and to cut the Gordian knot instead of untying it. Nevertheless, the account has to be settled sooner or later. In the end one has to admit that there are problems, which he simply cannot solve on his own resources. Such an admission has the advantage of being honest and truthful" (*Archetype And Symbol*, 1991, p. 111). What Jung meant was that even a strong person fails to grasp the contents of his unconscious sphere due to a peculiar mechanism of its functioning.

However, what's the use of knowing the archetypes of one's personal unconscious?

Jung's answered this as follows: "It's a tight passage, a narrow door, whose painful constriction no one is spared who goes down to the deep well. However, one must learn to know oneself in order to know who one is. For what comes after the door is, surprisingly enough, a boundless expanse full of unprecedented uncertainty, with apparently no inside and no outside, no above and no below, no here and no there, no mine and no yours, no good and no bad. It is the world of water, where all life floats above all" (*Archetype And Symbol*, 1991, p. 111).

The author of analytical psychology accurately describes "the way to himself": "One must descend to the water to cause the miracle of waters' survival." (Water is most common symbol of the unconscious. Jung defined water not as a metaphor, but as a life symbol existing in the dark of the soul.) "The breath of a spirit sliding over the dark water is terrifying like everything whose cause we do not know, since it is not ourselves. It hints at an unseen presence, the numen" (*Archetype And Symbol*, 1991, p. 108). [Jung borrowed the term "numinous" (lat. numen – the active power of the divine) from Rudolf Otto the author of "The Holy" (Otto, *Das Heilige*, Munchen, 1917), which describes the phenomenon of the divine experience as something both awesome and appealing, both fearful and attractive (*Archetype And Symbol*, 1991, p. 289).] This passage clearly shows the traditional approach of psychoanalysis, i.e. submerging into the depth of one's unconscious in search of one's soul.

When touching upon the subject of the numinous, many seek after a spirit in religion and mystics. In his work "Archetypes of the collective unconscious," Jung emphasized that the spirit comes to whomever he chooses and man is powerless to either find or take possession of him. "Neither human expectations nor the machinations of the will have given him life. It lives of itself, and a shudder runs through the man who thought that "spirit" was merely what he believes, what he makes himself, what is said in books, or what people talk about. However, when it happens spontaneously it is a spooky thing, and primitive fear seizes the naive mind" (*Archetype And Symbol*, 1991, p. 108). Jung stated that the elders of the Elgony tribe in Kenya gave him exactly the same description of the nocturnal god: "He comes to you," they said, "like a cold gust of wind, and you shudder, or he goes whistling round in the tall grass...We must surely go the way of the waters, which always tend downward, if we would raise up the treasure. When our natural inheritance has been dissipated, then the spirit too, as Heraclitus says, has descended from its fiery heights" (*Archetype And Symbol*, 1991, p. 108).

We have arrived at a conclusion that the way to get to know yourself, hard and long as it is, clearly leads to a spirit through learning one's soul and one's unconscious, i.e. through one's individual archetypes. It leads to the "dark mirror that reposes at its bottom" (*Archetype And Symbol*, 1991, p. 108).

However, there's an alternative way, and to use an allegory, it's "walking along the sunlit path laughing and cheering". Here, applying centuries long stock of the collective unconscious one might not need to worry about going deep into one's own "dark waters," trusting the power of one's intellect, the stability of one's collective values and society's support. And, as was mentioned before, the fee for that seems like not too high – it's either one's personality or one's lost soul. However, what is a human without his personal soul? By comparing with a plant rooted out of the soil and its subsequent dying, we can see the humanity's slow death in case the connection with his individual nature is lost. True modern man, especially a big city dweller surely can't have as much contact with nature as he had before. But this is beside the point since no matter where he lives he should still be in touch with nature through his psyche. Since a human is a product of nature, we feel his psyche should contain similar natural elements and phenomena that support his vital functions whether he is on a top floor of a skyscraper or orbiting the Earth in a space shuttle. His psyche, or his soul is what connects him to nature and only due to this connection not to some god or devil the human can live and function.

The interpretation of symbols is most important in getting to know the unconscious sphere since they express the very contents of the unconscious, the ones we, feel but can't define. Why? The symbol can be interpreted from different angles. Even though man appeared before all religions still a most intelligent person may be so naive as to believe that he was created by God, forgetting though that the only god before him and the one that could really create him was first and foremost nature itself. Therefore, his connection with nature could only be presented in the language of nature's analogs. All the rest including the origin of the world or man was later either created by human intellect or might have been the artifacts of culture.

One may wonder why all these religious ideas as to the creation of man, his purpose and his psyche have been so popular? Probably because real values have either been forgotten or profaned and replaced by artificial ones like soy protein substituted for meat in modern frankfurters. What happens then? This is how Jung elaborated on the problem: "According to anthropologists, in primitive societies with their spiritual values damaged by civilization, people often lost purpose in life, getting disorganized socially and corrupt morally. We are now facing the same situation. However, it never occurs to us what we really have lost. Our spiritual leaders are more concerned with protecting their social institutions than understanding the mystery of symbols" (*Archetype And Symbol*, 1991, p. 85). The situation seems to be quite trivial when religious and social institutions, caring so much about their influence on people's mind, hide from people most vital information as to who they really are. The fact of the matter is that all these institutions are mostly preoccupied not with the welfare of a concrete individual, but with that of a society as a whole. If a person in his attempt to learn of himself can cause problems for society, any such attempt will be blocked up by ideas coming out of culture, science or religion. And, this is exactly what's been happening. "Overly cultured" modern man slowly but surely withers like a plant rooted out of the soil, regularly watered by artificial bioactive growth stimulants in the form of various ideas and symbols of the collective unconscious, which society is trying to substitute for those of its personal unconscious. True, after feeling the ground slipping away from under his feet, modern man can still last for some time on artificial formulas of the collective ideas. However, what can really save him from imminent death is his return back to nature, not relocating to some suburban area "close to nature" as they were, but to his own inner nature, through getting to know himself and his own psyche.

All speculations presented in this article have been moving us to better understanding the contents of the unconscious stratum of human psyche, the ones which happen to be completely independent either from each other or racial or social factors. Even if we were to accept Jung's theory of "dividing man's psyche" we believe it would be his personal unconscious, which contains some very definite individual structure or system meant to define his personal psychological and physiological characteristics because the unconscious sphere of one's psyche directly controls many physiological reactions. The personal unconscious should be able to tell everything of one's personality not by examining it through its collective values, but in the light of its individual archetypes. It should be able to explain everything about a person with respect to his individual perception of the world and of himself, his purpose, his behavior both in the social and natural environment, his strategies and tactics, his way of thinking, his unconscious motivations, his needs, emotions, inclinations, preferences, meaning of his existence, etc.

After reviewing our both practical and theoretical research, and focusing on Jung's theory as to the influence of the collective unconscious on individual human psyche, it would be essential to examine this issue from the standpoint of the psychology of individual distinctions. It's obvious that without finding individual archetypal patterns as the basis for human personality, solving any individual or social-collective problems wouldn't be possible since any society consists of individuals constantly affecting one another.

References

Jung, C. G. (1991). A. M. Rutkevich (Ed.). *Arkhetip i simvol* [Archetype And Symbol]. Moscow: Renessans.

ARTICLE 2

CAN ARCHETYPAL IMAGES CONTAIN CHIMERAS?

Andrey Davydov and Olga Skorbatyuk

Translated by Kate Bazilevsky.

It is logical to suppose that the matrix of human psyche, which, in essence, determines *natura hominum*[1], and which for centuries scientists, philosophers, and mystics are looking for, is recorded by exclusively taking into account its transmission "through nations, centuries and cataclysms." According to researchers of antiquity, the human race is incredibly old: from tens of thousands of years to tens of millions. And since, as the ancient proverb states "Mother earth shook more than one civilization off its back," there must be that language, by the means of which all information about nature of human psychophysiology (that is—baseline information about human species) could be passed on from generation to generation, from civilization to civilization. By definition, only the language of natural forms could serve as such. Images of natural objects and phenomena are not only understandable by any inhabitant of the world, regardless of his or her nationality, religious views, level of education and intelligence and so on, but also cannot be destroyed as a result of wars, revolutions, catastrophes, and collapse of cultures, which have occurred in the history of humanity in huge quantities, destroying huge layers of information.

According to researchers, about 10,000 years before the Christian era, there was a period that lasted for several centuries, when any knowledge was suppressed, any information got destroyed, monuments were overthrown, and all material traces of disappeared civilizations were literally erased from the face of the earth. However, here and there huge constructions like the strange monoliths of the Easter Island remained, and to this day they are silent witnesses of lost arts, sciences, disappeared races. Researcher Manly P. Hall, in his book called The Secret Teachings Of All Ages: An Encyclopedic Outline Of Masonic, Hermetic, Qabbalistic and Rosicrucian Symbolical Philosophy, in our opinion, made a correct supposition: "Far-sighted were the initiates of antiquity. They realized that nations come and go, that empires rise and fall, and that the golden ages of art, science, and idealism are succeeded by the dark ages of superstition. With the needs of posterity foremost in mind, the sages of old went to inconceivable extremes to make certain that their knowledge should be preserved. They engraved it upon the face of mountains and concealed it within the measurements of colossal images..." (Hall, 2001, p. 37) Indeed, the sages were trying to preserve valuable knowledge from senseless destruction by savages, who are unable to comprehend the value of this information: "Their knowledge ... they hid within mythologies which the ignorant would perpetuate..."

In mythology, mandatory presence of certain natural elements in appearance, clothing and attributes of ancient gods and goddesses unambiguously indicates that "Nature is the source of nutrition for plant, animal, and human life" (Hall, 2001, p. 133). However, nature turned out to be not only the base of resources for the entire human life activity, but also the creator of a universal language, which can be passed on regardless of any kind of social cataclysms. Natural forms themselves turned out to be this ancient language, which does not belong to any one race or culture, and therefore is most suitable for transmission of information about the structure and mechanisms of functioning of human psychophysiology.

Despite many fundamental differences in theories, we consider ourselves followers of the great Swiss psychoanalyst, psychiatrist and founder of analytical psychology—Carl Gustav Jung. Following his footsteps, we also turned to mythology, symbolism and mysticism in search of answers to some of our questions. As it is known, the main function of mythology is that important information about a human and the world gets passed on through it from generation to generation. This factor specifically causes its presence in all cultures, widespread and extraordinary viability throughout the centuries: esoteric and philosophical, religious and ideological systems collapse, disappear, exeunt, get replaced, while mythology was, is and will be. Elements of myths are always present in the life of different nations: in a variety of titles, names, heraldry (which has taken its unicorns, fire-breathing panthers, dragons and many other creatures from these sources). Myths live everywhere: from the culture of samurai helmets to modern automobile design, from carnivals to armies of different countries around the world and components of everyday life of a person of any nationality, even computer icons also represent a certain variant of modern mythology. It is easy to trace that myths of all cultures in essence are identical and their various components in different cultures are identical since they represent the same characters: a dragon and a hydra, a wolf and a werewolf under different names in different nations carry the same symbolic meaning.

On the other hand, while mythology is constantly present in people's lives, its essence is still not understood: what do mythological stories convey, describe, what are they really about? Although more than enough of all sorts of interpretations exist, the question that interests psychologists is:

how can all that what is described in myths practically help the process of learning specific content of the unconscious of a particular person? We think that this question lies almost on the surface, as it is said—*veritatis simlex oratio est*[2]: the Egyptian and Greek mythology, the ancient Mysteries, shamanism, bestiary base on images of predatory animals or visions of our contemporaries in transpersonal states, in which people see themselves as animals—this indicates that people contain these animal forms within themselves. Such supposition has already been confirmed by our research, and it was uncovered how it is necessary to decrypt this or that animal image in order to obtain specific information about particular psychophysiological properties and qualities of the person being studied. However, the following question emerges: why in the ancient mystical and religious cults and practices, in myths and various psycho-techniques (ancient and modern) images of animals are often presented as chimeras? What is it that they are pointing to as far as learning of the content and mechanisms of functioning of human psyche? Therefore, we hypothesized that images of archetypal pattern can have chimerical images.

Chimeras (or so-called "composite creatures") are often not only characters of mythology and ancient cults. They are also widely presented in the ancient and medieval esoteric, mythological and even natural science literature, and are also found in the Mysteries. We have decided to start our study of chimeras specifically with the Mysteries. What are Mysteries? Using the modern language, such events could probably be described as "groups of deep self-knowledge and personal growth." Since "The Mysteries were therefore established for the purpose of unfolding the nature of a man according to certain fixed rules which, when faithfully followed, elevated the human consciousness to a point where it is capable of cognizing its own constitution and the true purpose of existence" (Hall, 2001, p. 241). However, such comparison would be incorrect for many reasons, which include not only the methodology of, but, first of all, results of such activities. Why? The thing is that the ancient really did have secret knowledge, which could qualitatively change human nature, advance it much further from the original point, while the modern psychology does not yet have such methods. The fact that the Mysteries provided great and eternal truths was confirmed by many outstanding minds of antiquity: "Would Pindar, Plato, Cicero, Epictetus, have spoken of them with such admiration, if the hierophant had satisfied himself with loudly proclaiming his own opinions, or those of his order?" (Hall, 2001, p. 71) It is possible that after gaining knowledge about the existence of individual archetypal patterns, modern psychologists will be able to overcome this situation.

It is not without reason that philosophy itself, as a "science of wisdom," which also emerged from esotericism in the form of Mysteries of antiquity, was not separated from them until the Mysteries have faded. Manly P. Hall wrote: "Hence he who would fathom the depths of philosophic thought must familiarize himself with the teachings of those initiated priests designated as the first custodians of divine revelation. The Mysteries claimed to be the guardians of a transcendental knowledge so profound as to be incomprehensible save to the most exalted intellect…" (Hall, 2001, p. 36) Justification of statements about having Universal Wisdom allowed these institutions of the past to reveal depths of their secret teachings to great and famed philosophers of antiquity, such as Thales, Pythagoras or Plato. Universal Wisdom by definition could be given only to those people who could understand philosophical concepts. It appears that they were considered "chosen" based on this criterion specifically. When discussing philosophy, it is also necessary to recall the fact that even though Hellenes have shown themselves as a very capable nation in this sense, the invention of the "mother of sciences" does not belong to them. "Although some of the Grecians," wrote Thomas Stanley, "have challenged to their nation the original of philosophy, yet the more learned of them have acknowledged to [to be] derived from the East." "The question may legitimately be propounded: If these ancient mystical institutions were of such "great pith and moment," why is so little information now available concerning them and the arcana they claimed to possess?" – questioned Hall, and then provided a response – "The answer is simple enough: The Mysteries were secret societies, binding their initiates to inviolable secrecy (which is why another famous ancient philosopher Socrates refused to be initiated into the Mysteries – our note), and avenging with death the betrayal of their sacred trusts (however, Plato managed to avoid it, even though he was accused of divulging some elements of teachings in his philosophical writings – our note)." As we now understand, these secret societies, schools were true inspirers of various doctrines spread by the ancient philosophers, while the source itself (the Mysteries) was never revealed to the profane (uninitiated). It is not surprising that over time teachings of philosophers that were based on doctrines, which really belonged to these secret institutions, began to be considered inseparable from and together with the names of their distributors, and their true source—the Mysteries—was forgotten and began to be completely ignored (Hall, 2001, pp. 36-37).

In his work devoted to attempts to interpret the Secret teachings hidden behind the rituals, allegories and mysteries of all time, Manly P. Hall wrote that in the ancient rituals of initiation of the Persian, Greek and Egyptian Mysteries "priests disguised themselves as composite creatures, thereby symbolizing different aspects of human consciousness" (Hall, 2001, p. 274). In addition, he noted: "The Egyptian sphinx, the Greek centaur, and the Assyrian man-bull have much in common. All are composite creatures combining human and animal members…" (Hall, 2001, p. 287) What other chimerical creatures are found in the descriptions of ancient cults? Firstly, of course, we come across various deities. For example, in the Mithraic Mysteries, Mithra as Kronos is depicted as having a human body, four wings and a lion's head, his body from toes to head is entwined by a large snake, which holds a key over his head. The famous Abraxas, a Gnostic deity is depicted as having a human body, two snakes instead of feet (symbolizing Nous (Mind) and Logos (Word), which means inner feeling and quick understanding – our note) and a Rooster's head, which represents foresight and vigilance. Also found in the mysteries is the ibis-headed Egyptian Thoth, Lord of the divine Books and scribe of the Company of the Gods, who is also, apparently, the famous Hermes Mercury Trismegistus Thrice-Greatest often depicted as having a dog's head[3]. Besides the specified, there are many other Egyptian deities that are chimeras, and this is easily verifiable by looking at the ancient monuments of the Egyptian culture; for example, the famous Bembine Table of Isis[4].

In addition to gods, other chimerical creatures were involved in the Mysteries. We will name a few as examples. in regard to one of these creatures M. Hall wrote: "The unicorn, or monoceros was a most curious creation of the ancient initiates. It is described by Thomas Boreman as "a beast, which though doubted of by many writers, yet is by others thus described: He has but one horn, and that an exceedingly rich one, growing out of the middle of his forehead. His head resembles an hart's, his feet an elephant's, his tail a boar's, and the rest of his body an horse. The horn is about a foot and half in length. His voice is like the lowing of an ox. His mane and hair are of yellowish colour. His horn is as hard as iron, and as rough as any file, twisted or curled, like a flaming sword; very straight, sharp, and every where black, excepting the point. Great virtues are attributed to it, in expelling of poison and curing of several diseases. He is not a beast of prey." (See Redgrove's *Bygone Beliefs*.)" (Hall, 2001, p. 300) Another chimerical being involved in the Mysteries was the Manticore (mantichora). Ctesias describes it as having a body of flame-color, shape like the body of a lion, having three rows of teeth, a human head and ears, blue eyes, a tail with thorns and spikes on its end like a scorpion, and having a voice "like the blare of trumpets" (Hall, 2001, pp. 279-280).

Manly P. Hall wrote about Manticore: "This synthetic quadruped ambled into mediaeval works on natural history, but, though seriously considered, had never been seen..." (Hall, 2001, p. 280) Impossibility to see Manticore was explained, allegedly, according to the source, by that it lived in inaccessible places or that it was difficult to see. Here, two facts attracted our attention: firstly, the animal, which based on the description is difficult to attribute to any category other than the category of fantastical, despite the fact that it "had never been seen," nevertheless was present in works on natural history! And even, as the author asserted, was "seriously considered." Secondly, places of habitat of this supposedly existing, and for that reason seriously considered Manticore are described by rather vague landmarks. This caused the supposition that the animal-chimera itself and it "existing in the invisible world of Nature" (Hall, 2001, p. 303) clearly indicate that this refers to an animal image in the form of a chimera, really existing in the unconscious of a human. There is also a direct indication by the researcher of symbolism in favor of this supposition: "...in the Mysteries all signify the composite nature of man..." (Hall, 2001, p. 287)

Hence, "non-existing animals," involved in the Mysteries on equal rights as real ones, and which in them became a reality, as M. Hall wrote: "signified the manifold of man's nature" (Hall, 2001, p. 274), complex chimerical images of the human unconscious. Apparently, this was one of the most important riddles of the Mysteries, which revealed the secret of the human nature, the human soul. This fact quite fully explains the interest of the ancient scholars and mystics in the Mysteries, and these people cannot be considered "empty-headed", "foolish" or ignorant.

Fantastic animals along with real can be found in works of Aristotle and Pliny, Arrian and Pausanias, and Thales of Miletus. By the way, the latter made an interesting statement, which illustrates "one of the first views in science about a human as an animal": "Species lies in genus, as a human in an animal" (Terentev-Katansky, 2004, p. 13). The fact remains: according to writings of researchers, the difference between "a fantastic animal" and "an existing (real) animal" was highly conditional in both the ancient times and the Middle Ages. Works of Isidore of Seville, Ambrose's Hexameron, as well as Rabanus' books can be considered as a transitional form from books on natural science to half-fantastic, half-real genre—bestiary that springs from the famous Physiologus, which includes descriptions of over two hundred animals, birds, reptiles and fishes, with the addition of plants and stones in some lists. According to researchers, "a peculiarity of bestiaries was that they gave sacred, ethical and didactic meaning to an animal, while the image of an animal held a purely informative, natural-scientific value" (Terentev-Katansky, 2004, p. 14). Many of bestiaries' characters are found in books of medieval researchers; for example, a number of such creatures are in the work of Villard de Honnecourt (1235), there are also images of wild animals and humans in inscriptions and drawings on a map made in Lüneburg in the year 1248. Such major authorities as Olaus Magnus (1490-1557), monk Thevet (1516-1590), Aldrovandi (1522-1605), Conrad Gessner (1516-1565), and Athanasius Kircher (1602-1680) took a lot from bestiaries. The most famous authors of bestiaries in medieval Europe can be considered Philippe de Thaon (XII century), Richard de Fournival (XIII century), travelers Jacques de Vitry (1170-1240), Marco Polo (1254-1324), who contributed numerous wonders of the East to bestiaries. However, according to a researcher of Chinese culture A. P. Terentev-Katansky, it is likely they got there before them (Terentev-Katansky, 2004, p. 15-16).

According to researchers-sinologists, very interesting examples of books about fantastic creatures exist in China, and, by the way, they are literally filled with chimeras. Among all the literature of this kind, Shan Hai Jing (translated as Catalog of Mountains and Seas) occupies the first place. It is a book the date of which is still unknown and ranges from XXI to III centuries BC. The Catalog is as if divided into two books. The first of these two books is called the Catalog Of (Five Inmost) Mountains, and its main content is description of geographical zones with elements of zoogeography, botanical geography, folk medicine, folk beliefs and mythology. The second book, called the Catalog Of Seas, includes information in the form of lapidary lists and separate fragments about lands and people, which, according to Qin-Han's sinocentric concept were located on outlying lands and outside of the inhabited world. "This information, – stated the author of the translation (into Russian) of this most interesting historical monument of the Chinese culture, E. M. Yanshina, – are mostly of mythological character—it contains fragments of myths about ancestors, heroes and their genealogy" (Yanshina, 2004, p. 8). The book consists of descriptions of amazing animals, birds, plants, spirits, fantastic people, and contains a huge quantity of various chimeras. According to legend, Bo Yi (an assistant to Yu the Great—the mythical organizer of the earth and eliminator of the great flood) engraved images for this ancient book on nine ritual vessels. These vessels were allegedly later on lost and rediscovered during the Early Han (II century BC to beginning of I century AD). The text of the Catalog that we know today was, allegedly, written off of these tripods (Yanshina, 2004, p. 10). There were other books also presented in the Chinese culture, which described various fantastic creatures. The Erya, for example, is the same edition, which Gould used for his book Mythical Monsters, published in London in 1886. Among Chinese bestiaries it is possible to also mention Bencao Gangmu (*Compendium of* Materia Medica), the work of the great physician of the XVI century Li Shih-chen, which in addition to the section that describes plants, also has a section on animals, as well as many others.

It should be noted that mythology of different nations touches upon the same theme in one form or another. According to their beliefs, unusual creatures once inhabited earth: outlandish animals, birds and strange humanlike creatures. Specifically because of mythology, we have become used to complex "composite creatures"—chimeras in description of various mythological events. It seems that really no one is surprise by a Centaur, whose upper body is that of a human and lower—of a horse, or a hippocamp—part horse, part dolphin. We have calmly accepted a Phoenix, also a chimera, often depicted with a human head, neck of a snake, tail of a fish, crane's forehead, small beard like that of a male Mandarin duck, dragon's patterns, turtle's back, swallow's goiter, and beak of a rooster; or the famous Dragon, about which the following was written: "During the Song Dynasty formed a teaching about "eight similarities" of a dragon: antlers of a deer, head of a camel, eyes of a spirit (in some sources eyes of a rabbit), neck of a snake, scales of a carp, claws of an eagle, paws of a tiger, ears of a cow" (Terentev-Katansky, 2004, p. 28). Mythology prepares our perception for that what we cannot know based on our empirical experience because we can actually come across these images in the content of our unconscious, which erupt "outward": in dreams, states of trance, etc.

In light of the topic being addressed, there is an interesting fact of existence of a legend, according to which long before humans existed there was an entire race of "composite creatures." "The temples of antiquity – stated Manley P. Hall – preserved their own historical records and possessed information concerning the prehistoric world that has never been revealed to the uninitiated. According to these records, the human race evolved from a species of creature that partook somewhat of the nature of an amphibian, for at that time primitive man had the gills of a fish and was partly covered with scales" (Hall, 2001, pp. 274-275). At one time, the consequence of such "theory of man's origin in water" was the point of view that a fish is the primogenitor of humanity. According to M. Hall, this explains worship of a fish by the Phoenicians, Brahmins, and Chaldeans. Such legends have led to that the ancient peoples began to associate the image of a fish with World Saviors. Here are some of them: the Vishnu god, who was pulled from a fish's mouth; the Isis goddess, who was depicted with a fish in her headpiece and suckling her son Horus; Oannes, the Chaldean Savior adopted from the Brahmins, was depicted with head and body of a fish (also a chimera – our note), from which a human form appeared in different places. By the way, a fish was the first monogram of Christ, adopted by the early Christian church as a symbol of Christ (Hall, 2001, p. 246). A modern educated person will perceive this information differently, for example, as another tale, or as a direct indication of either evolution of humanity in light of Darwin's theory of human's origin, or as an indication of the fact that the human embryo goes through a similar stage in the course of prenatal development. In the light of new knowledge about the existence of archetypal images of the unconscious, which can be chimerical, it is possible to consider this in a different way. Knowing that matrix of the entire human psychophysiology, the basis of psyche is a set of archetypal images, which we called an individual archetypal pattern of a person, it is possible to suppose that this is exactly what such stories are about.

Information from mythological and esoteric sources indicates that archetypes existed prior to appearance of humanity, from which (as from pieces of the Lego construction set) archetypal patterns, individual models of psyche were created. Archetypal pattern itself is a chimerical formation. Therefore, a chimera serves as the principle of certain relation and interrelation between parts that make up one whole in the human psyche. As we have seen, there is every reason to suppose that animal images of individual patterns can actually be chimeras, and results of our research confirmed this supposition. Consider the Sphinx, which, according to researchers, not only silently sits at the gate of the temple guarding the

Mysteries and does not allow passage to the profane, but also represents the secret doctrine itself. We supposed that the Sphinx is a formula, by which information about the nature of a human is recorded in his psyche. The Sphinx represents several important principles that relate to structure of the human psyche. The first of which is that the soul (psyche) of a human is one whole, but it consists of constituent parts—natural images that form an archetypal pattern. The second important principle is that the Sphinx is a chimera; therefore, images of individual archetypal pattern of a person really can have chimerical images. The third principle is that under the external human image (the Sphinx has a human's face), a human contains natural forms within himself (including animal forms, as the body of the Sphinx is a chimera—a winged lion). There is also a fourth principle represented by the Sphinx. According to Hall, the Sphinx, "...placed between man and his divine possibilities..." really represents individual archetypal pattern in the form of chimeras (Hall, 2001, p. 287). Being in the unconscious and remaining unknown, it does not allow a person to advance to qualitatively new levels of development that transform his nature and open up entirely different abilities. Many talents and abilities remain unactivated throughout life of a person who does not know the content of his unconscious, which now a person can know through learning images of his individual archetypal pattern.

"To the student of the secret wisdom, – wrote Hall in regard to chimeras, – these composite animals and birds simply represent various forces working in the invisible worlds" (Hall, 2001, p. 274). The problem is that there is no specification as to which invisible worlds are being discussed. In our opinion, the discussion is about the unconscious, which is the only one of the worlds that a person really does not see, but always feels its presence within himself. In addition, a person can, even has to gain knowledge of the world of his unconscious. The unconscious itself constantly pushes him to do so. In psychology, this phenomenon has been well studied, is known to all and is no longer questioned. Probably understanding of such things and similar associations have not been used by researchers of psyche because, due to their own ambitions, they had no idea of the abovementioned "individual projects" that already exist within the human species and their actual constituents. Mythology also has not been considered as a specific description of the unconscious, where outlandish chimerical animals are not just characters of a tale, but are chimerical images that make up individual archetypal complexes. However, apparently, this specifically is the essence of the Mysteries—gain of knowledge of self, one's nature, which took place in the Mysteries through a variety of sacred actions with chimerical images that represented the human unconscious. This specifically really led those people who participated in the Mysteries to "...cognize its own constitution and the true purpose of existence" (Hall, 2001, p. 241), which actually took place not through some "magical actions," but through learning images of their unconscious, which, once again, is a phenomenon widely known in psychoanalysis.

Therefore, it turns out that, in a certain sense, the emergence of symbolism has played an evil joke on humanity. For it is due to efforts of the ancient esotericists, priests to make some of their knowledge about the structure of the human psyche, of human nature maximally secret, giving them an enormous, sometimes unlimited power over the people, that all of the most important information was encrypted in symbols. Therefore, it turns out that by symbols the initiated not only "...sought to communicate to each other those thoughts which transcend the limitation of language," and "...rejecting man-conceived dialects as inadequate and unworthy to perpetuate divine ideas, the Mysteries thus chose symbolism as a far more ingenious and ideal method of preserving their transcendental knowledge." (Hall, 2001, p. 37) But, it turned out that a symbol became, according to Hall's expression: "the tomb of an eternal verity." However, as always, this is only one side of the coin. Since the sages were absolutely right in that there are things that should not fall into the hands of ignorant people. This really should have not turned the vast majority of researchers away from the study of symbols, but on the contrary should have stimulated them to improve their educational level and grow their consciousness in order to gain the ability to understand this complex language. However, this did not happen, and that is why symbolism turned out to be too tough for most researchers. At the same time it must be said that symbolism accomplished its mission—to save the secret knowledge from the ignorant, that is, people who are unable or do not want to put any effort into obtaining it. And, if there is a code, according to which, as Hall wrote, "every visible object, every abstract thought, every emotional reaction is but the symbol of an eternal principle," then there exists a possibility to decode it—"through education ... man will learn to release living truths from their lifeless coverings" (Hall, 2001, p. 258).

In order to fully understand the stand of mystics, we present an excerpt from The Secret Teachings Of All Ages, which explains a lot:

> "*When confronted with a problem* – states the researcher of symbolic philosophy, – *involving the use of reasoning faculties, individuals of strong intellect keep their poise, and seek to reach a solution by obtaining facts bearing upon the question. Those of immature mentality, on the other hand, when similarly confronted, are overwhelmed. While the former may be qualified to solve the riddle of their own destiny, the latter must be let like a flock of sheep and taught in simple language. They depend almost entirely upon the ministrations of the shepherd. The Apostle Paul said that these little ones must be feed with milk, but that meat is the food of*

> *strong men. Thoughtlessness is almost synonymous with childishness, while thoughtfulness is symbolic of maturity.*
>
> *There are, however, but few mature minds in the world; and thus it was that the philosophic-religious doctrines of the pagans were divided to meet the needs of these two fundamental groups of human intellect—one philosophic, the others incapable of appreciating the deeper mysteries of life. To the discerning few were revealed the esoteric, or spiritual, teachings, while the unqualified many received only the literal, or exoteric, interpretations. In order to make simple the great truths of Nature and the abstract principles of natural law, the vital forces of the universe were personified, becoming the gods and goddesses of the ancient mythologies. While the ignorant multitudes brought their offerings to the altars of Priapus and Pan (deities representing the procreative energies)* (which also appeared in the image of chimeras – our note)*, the wise recognized in these marble statues only symbolic concretions of great abstract truths*" (Hall, 2001, p. 38).

Every nation has not only a state religion, which is the "milk" for the "little ones" with its endless interpretations that really are necessary in order to make "the great truths" simple for the ignorant. Religion differs from the Mysteries in that it calls into its ranks and accepts into its fold everyone. The Pagan mysteries countered the Christian church in that they demanded virtue and integrity as a prerequisite to salvation through initiation, while the church was ready to accept into their fold sinners, and fools, and beggars, and paupers, and criminals, moreover, promising "Kingdom of God" to all of them. This was contrary to the principles of acceptance into the secret esoteric societies. In addition, a religious person worships visible things as symbols of the invisible, omnipresent, indescribable force, about which he knows nothing. Thomas Taylor wrote: "Man is a naturally religious animal" (Hall, 2001, p. 40), referring to the fact that from the "dawning" of his consciousness, he always sought, to deify material, visible things in one way or another. [By the way, is this not what science is engaged in to the present day, and, in the words of Eliphas Levi, frantically "recedes that it may advance the further?" (Hall, 2001, p. 43)] However, there is another knowledge—esoteric, which is the abovementioned "meat" for "strong men" (who are "able to chew on their own" – our note), that is to understand complex things, to which at all times (most likely to this day) only the intellectual, philosophical elite had access, and, of course, not all were admitted, but only those, who by their own qualities were somehow capable of getting there.

The main criterion of truth of this or that way, as it seems to us, is the will of each person to decide to what extent he can allow himself be truly free.

However, it is important to always remember that this directly depends on the level of his efforts directed to correct understanding of the world and himself in this world, which, by definition, is not possible without his knowledge of himself. As a "...man, gazing into the mirror of Nature and accepting as his real self the senseless clay that he sees reflected (meaning that for Eleusinian philosophers—birth in the physical world was death, and the only true birth was the birth of own soul through self-cognition and subsequent transformation of his consciousness – our note) loses the opportunity – wrote Manly Hall – afforded by physical life to unfold his immortal, invisible Self" (Hall, 2001, p. 70). Indeed, as it is clearly shown in esoteric, as well as philosophical and religious teachings: "Like Prometheus chained to the top of Mount Caucasus, man's higher nature is chained to his inadequate personality" (Hall, 2001, p. 82).

Therefore, the ancient sages were right: symbolism is not just an ideal way to preserve the secret knowledge. As M. Hall justly pointed out: "...in fact, it is the language not only of mysticism and philosophy but of all Nature, for every law and power active in universal procedure is manifested to the limited sense perceptions of man through the medium of symbol" (Hall, 2001, p. 37). Therefore, in order to really understand the content of that knowledge about human nature and the world around it is necessary, from our point of view, to say goodbye forever to the illusion that somewhere in the ancient records the "great secret" about the nature of the soul will be described in "simple human language": "Hence, he who seeks to unveil the secret doctrine of antiquity must search for that doctrine not upon the open pages of books which might fall into the hands of the unworthy but in the place where it was originally concealed." And, this place is a symbol, which should be considered as a formula. The initiated, who really knew the secrets about a human and the world, "departing, ... left their formulae that others, too, might attain to understanding," and it is not worth it, from our point of view, to be like profanes of the past and perceive all information from the mythology and mysticism literally. A good illustration of this is the whole history of symbolism, as well as the current situation in the sphere of human knowledge: "Today men gaze with awe and reverence upon the mighty of Memnons sanding alone on the sands of Egypt, or upon the strange terraced pyramids of Palanque. Mute testimonies these are of the lost arts and sciences of antiquity; and concealed their wisdom must remain until this race has learned to read the universal language—SYMBOLISM." Because it turns out that symbolism and mysticism in general, in our opinion, should not be underestimated. It is known that a mystic is a person who turns his attention to all the invisible and hidden, the word "mistes," as it is known, means "s/he who sees through the fog." Therefore, mysticism, as a way of cognizing the world and the area of human activity, also provides science with real abilities to access the so-called "higher truths," one of which, surely, is knowledge of the human soul, that is—knowledge of structure and mechanisms of functioning of human psyche.

To conclude the discussion about symbolism, we would like to uncover another way, as always known in antiquity, but undeservedly forgotten: "The sages of old studied living things to a point of realization that God is most perfectly understood through a knowledge of His supreme handiwork—animate and inanimate Nature" (Hall, 2001, p. 273). Therefore, it may be that the main merit of C. G. Jung (who studied symbolism a lot and had an encyclopedic knowledge in this area, but who has failed to find an answer to the question of how to translate a particular symbol into specific properties of human personality) can be considered that he marked the vector of subsequent research.

To answer the posed question about whether archetypal images can contain chimeras, as a result of research we can answer this positively. A different matter is that chimeras do not contain any images except natural, so a chimera can be considered as a kind of encoding, which, instead of symbolism that is not understood by most researchers, carries in itself the universal language of natural forms.

References

Hall, M. P. (1928). *The Secret Teachings Of All Ages: An Encyclopedic Outline Of Masonic, Hermetic, Qabbalistic and Rosicrucian Symbolical Philosophy*. San Francisco: H.S. Crocker Company, Inc. Scanned November 2001, by Hare, J. Retrieved September 10, 2014, from http://www.framsteget.net/gratis/TheSecretTeachingsOfAllAges.pdf.

Ezhov, V. V. (2003). *Mify Drevnego Kitaya* [Myths of Ancient China]. (Ill. *Myths of Nations of the World*) Moscow: Astrel.

Yanshina, E. M. (Trans.). (2004). *Katalog Gor I Morej (Shan Hai Tszin)* [Catalog of Mountains and Seas (Shan Hai Jing)]. (2nd ed.) Moscow: Ripol Classic.

Terentev-Katansky, A. (2004). *Illyustratsii k kitayskomu bestiariyu: Mifologicheskiye zhivotnyye drevnego Kitaya* [Illustrations to Chinese Bestiary: Mythological Animals of Ancient China]. St. Petersburg: Forma T.

Footnotes

[1] (Lat.) nature of human beings; human nature

[2] (Lat.) check by simple

[3] Hermes Trismegistus became famous and a part of pantheon of gods because he was able to persuade the Dragon Poimandres (deciphered as Mind of the Universe) to reveal to him the secret of nature of the Universe and the essence of the gods. The Dragon's answer is of interest, as it lies in the area of the topic that we are examining: "Before the visible universe was formed its mold was cast. This mold was called the *Archetype*, and this Archetype was in the Supreme mind long before the process of creation began. Beholding the Archetypes, the Supreme Mind became enamored with Its own thought; so, taking the Word as a mighty hammer, It gouged out caverns in primordial space and cast the form of the spheres in the Archetypal mold, at the same time sowing in the fashioned bodies the seeds of living things" (Hall, 2001, p. 103). By the way, the Mysteries taught that the Universal Life is personified in the form of a dragon.

[4] There is an interesting mention of this famous table in Thomas Taylor's manuscript: "Plato was initiated into the 'Greater Mysteries' at the age of 49. The initiation took place in one of the subterranean halls of the Great Pyramid in Egypt. The ISIAC TABLE formed the altar, before which the Divine Plato stood and received that which was always his, but which the ceremony of the Mysteries enkindled and brought from its dormant state" (Hall, 2001, p. 170).

ARTICLE 3

SOCIETY AS A COMMUNITY OF MANIPULATORS AND THEIR SUBJECTS

Andrey Davydov and Olga Skorbatyuk

Translated by Kate Bazilevsky.

The following statement can be considered the main hypothesis that formed the basis of this article: all types of interpersonal relationships in the human community are built entirely on manipulation as a principle. Any kinds of contacts and attachments between people (friendship, romance, business, partnership, parent-child relationships) are based on manipulation of each other as an unconscious motivation. Consequently, the human community lives, develops, functions and is regulated on the basis of manipulation. And, if manipulation of one another by individuals is an effective way to quickly and qualitatively satisfy any kind of their needs in society, then it is fair to consider society as a community of manipulators and their subjects.

Let's examine this statement. It must be noted that our research is carried out in line with psychoanalysis. We do not study the conscious part of the psyche. We study the unconscious, and therefore, when we consider the question of motivation, we are interested solely in unconscious motivation. It is known that all attempts of an individual who is being studied to declare anything about himself or herself, as a rule, are not considered by a psychologist of psychodynamic orientation, as these declarations have no relation to the actual situation whatsoever. A story told by a person (a client, a person being studied, a test subject) about himself, his beliefs, perceptions, as well as motivations have a purely self-presentation character, and nothing more.

The concept of "motivation" exists in psychology and, as practice shows, it corresponds to a much greater extent to that what any types of self-presentation actually represent. After all, a human, as a "social animal" is forced to live by the laws of society, is raised since childhood in a manner to conform to these laws, and learns to hide his or her real motives. This is the essence of upbringing: to one way or another socialize an individual, to teach him or her social demands, without which social interactions become impossible. From our point of view, upbringing does not have any other functions because a human is already born with a certain set of psychophysiological properties and personality qualities recorded in his or her individual archetypal pattern, which cannot be changed neither by parents nor by caregivers. Therefore, we focus only on the study of unconscious motivations, as the true cause of human activity. And, when discussing true motivations that arise from the unconscious, we can state that they are fully determined by individual structure of psyche of a person. All true motives of an individual are fully determined by the content of his unconscious: that is—by individual archetypal pattern as the basis of an individual, where all needs and predilections of the individual are contained.

However, going back to the question of upbringing as a way of socialization of an individual, it is possible to state the following: if the necessary degree of socialization as a result of actions during upbringing by family and school for whatever reason has not been reached, then a person has difficulties with interpersonal interactions throughout his life. And, these difficulties not only lead to antisocial actions, but are also causes of internal problems of a human. Why? What is the basis of the phenomenon that an individual who is antisocial according to an innate quality, or is poorly socialized in the process of upbringing, can be literally torn apart by internal conflicts all his life, due to the fact that he does not "fit in" with society? Representatives of psychoanalysis have described such kinds of intrapersonal problems related to society and its impact on a person well enough, and therefore we will not continue with discussion of this issue. Here, we became interested in a different matter: why a person with antisocial orientation (as a property of his nature that is inherent in his individual archetypal pattern) begins to experience internal conflicts due to impossibility of harmonious interaction with other people? Why even an antisocial individual begins to feel uncomfortable due to lack of communication? Why is so important for a person to harmoniously integrate into the social mechanism, and what is the goal of this "integration?" Why community of his own kind is necessary for any individual (regardless of degree of his socialization), and is this society truly made up of people who are alike?

We have decided to clarify this question in the range of the statement that "to live in society and be free from society is impossible." However, we consider the relation "individual-society" based on understanding that every person living in the society and having a firm dependence on it, above all is an individual who is guided in his actions only by those values, which are inherent in this individual. This view is somewhat contrary to the customary point of view in regard to existence of universal human values. However, we consider universal human values as certain standards invented by people as a unifying element and nothing more. In reality, analysis of attitudes of different people towards so-called universal human values clearly shows that in essence none of the declared "values" are universal. Rather, it is common for a particular group of individuals, but that does not mean the totality of its perception as a value. Kindness, faith, love, children, work, creativity, health, learning, freedom, equality, brotherhood, even human life are not valuable to all the people. This is obvious simply due to the fact that there are murderers and suiciders, to whom neither their own lives, nor lives of others hold any value. And, most notably, they have lived, live, and will live like this, regardless of whether it is condemned by society or not. These people (and, by the way, there is a huge percentage of them in the society) can be persecuted, and they can hide or not hide their unconformities with social rules, norms and ideals, but even the most severe sanctions by authorities, condemnation, epitaphs or exile cannot eradicate these manifestations in people. Why? The answer can be quite simple: such is the nature of these people, and it cannot be changed by anyone except themselves. Therefore, it is possible to suppose that only power obtained as a result of manipulative actions can be a universal goal that people aspire to one way or another; power, with the help of which it is possible to stabilize obtainment from society of necessary components for comfortable psychophysiological state. This aspiration can be valued or despised, denied or accepted, but no one has yet been able to live entirely without this aspiration. Therefore, if it is so important to humanity to have at least something socially universal and common—then, this aspiration can successfully be that.

An individual goes to the human community with his needs in order to satisfy them—these are objective needs of his psychophysiology for something that lies outside of him, but constitutes a necessary condition for his normal functioning. Therefore, he is forced to accept rules of society, or he risks becoming an outcast. By nature not being that what he must be in order to be accepted by others, each individual is forced to wear this or that socially acceptable mask, and this is also a well-described phenomenon in psychology. However, this mask eventually forms changes in his self-perception within him, the so-called "SELF-image." It begins to seem to him that he really feels, really wants, really has those qualities, which he is forced to declare only in order to be socially acceptable, accepted, and understood by other people. That is exactly why, as researchers who study the unconscious, we can confirm the already familiar: a cover, figuratively speaking, is not yet a book. However, presence of information about individual archetypal patterns allows for an explanation of relations in society from other standpoints. However, before we get into that, we would like to state our principled approach to human psychophysiology.

When discussing archetypal pattern, it is possible to call it differently. For example: "individual psychophysiological algorithm", "individual structure of psyche" or simply "a program." It follows that since the entire individuality of a person as characteristics of psyche and physiology from birth is recorded in such matrix, it is possible to make the following thesis: a human is a bio-form, which has a program recorded in his unconscious, on the basis of which he operates analogously to all other bio-forms (animals, plants, etc.). This statement contradicts centuries of attempts to discuss freedom and will of a human. However, with presence (detection) of individual archetypal patterns in the unconscious of a person, access to which allows description of all his characteristics, a human, in essence, is a bio-robot programmed by nature. In this connection, the focus also changes: that what was originally called "a human" is a bio-robot.

Since this turned out to be so, it is absolutely obvious that a human, like any complex system, must be regulated and self-regulate. It would be unwise to assume that nature did not implant in a human—bio-robot any mechanisms for automatic, natural regulation of his psychophysiology. In search of mechanisms of regulation and self-regulation of a system called "a human," we found some built-in mechanisms in the unconscious and called them human manipulation modes. These regimes, like buttons on a remote control, can fully regulate all human psychophysiology, acting on psyche and physicality as two completely inseparable values in the life process.

Manipulation modes are three individual archetypal patterns that differ in orientation, and due to psychophysiological characteristics enclosed in them can encourage a person to act, calm him down, give him a sense of balance and harmony within himself, provide a deep sense of pleasure and satisfaction, and so on. It was the discovery of manipulation modes as mechanisms of self-regulation of a human that provided the answer to our question: what does a person really look for in the environment? We suggested: could it be that his manipulation modes, which are necessary in order to meet all of his psychological and physiological needs, are that what an individual tries to get from other people? Is this why he requires friends, family, communication with relatives, acquaintances? Answers to these questions were positive. It turned out that a human needs society as his environment mainly to satisfy his natural needs by getting his manipulation modes, which stabilize his psychophysiology, through communication with other people.

An important point to understand is the following: manipulation modes, being types of correctors of human behavior, mainly are regulators of functioning of a person's entire psychophysiology. They directly affect the activity of a person, can raise or lower it almost to "zero," effect his state of health and mood, improve or worsen them, and so on. Therefore, it is absolutely logical for a person to search for those people in the environment, who, through their existing individual archetypal pattern, transmit their physical and psychological characteristics in an interpersonal contact, and with this provide the needed "charge" to the other person. From our point of view, this explains the inability to be free from society.

Then, how are interactions constructed in the society of people who are considered to be, but by nature are not like each other? Simple enough, but, in our opinion, not constructive—there are rules that exist in society as requirements for reactions and actions of separate individuals, which a person must comply with if he is in society. There is a huge mass of requirements and expectations in regard to what an individual should be or appear to be (because factually it does not matter) in order to be accepted and demanded by others. If an individual is outside of these limits in some way, then the society has two main methods of getting rid of him. The first method can be called mechanical, when an individual gets physically removed. The second method can be called "virtual" because the wrongdoer or simply a non-standard individual physically present in society is factually isolated from it by various means. Thus, mortification of an individual, his forced isolation in prisons or psychiatric institutions fall under the first variant. Stigmatization of a person by various diagnoses such as "insane", "addict", "criminal"—hinder or even close the possibility for many, if not all social contacts and fall under the second variant. However, as we have already mentioned, no sanctions are followed by qualitative consequences: people die and new addicts, murderers, and "abnormal" people are born. Why? Now we have the answer to such questions—this occurs because a human, as a bio-robot, acts according to that "program," which nature implanted in him.

It is necessary to note that no religion in the world was able to "kill the animal" in majority of people belonging to a particular congregation, even fanatical believers. Or rather it was possible to "kill the animal" in a human only together with the person, in which it was "enclosed." However, such proposals were put forward and implemented in various religious confessions and esoteric teachings instead of allowing a person to live a "natural life of an animal" at the primary level, after explaining to him the fact that one simply needs to grow to become a human; that is—allowing him to calmly and naturally pass through this state. Since, as practice shows, only after successfully passing through this and subsequent steps, a person who has really set a goal for himself to go beyond the "animal-plant" state of existence can reach the state of likeness of the Creator. No person will succeed by jumping over evolutionary stages of development of consciousness in this process, and it is really unnecessary. However, how to let a person be his natural self, knowing that some of his individual characteristics can be harmful to others? It is surprising, but by following his individual animal nature, each person lives in his own strictly allotted ecological niche. Acting in accordance with his natural functioning, he is a natural segment of the biosphere, stays in his place with pleasure and without interference with others. This perspective is somewhat different from the modern process, in which all people in society are expected to have a unified way of thinking and acting. What in reality is a human community when considered from other perspectives?

Unsatisfactory and unnatural state of thing in society, in our opinion, occurs not only because of certain unconscious mechanisms that make it difficult for a person to understand and accept the seemingly obvious thesis—people are different, but also because few people perceive this thesis as a fact. Moreover, there are many social mechanisms, which grade these differences and hide them. For example, the already mentioned "universal human values." What also compounds the situation is that, firstly, all people are biologically identical, and secondly, the consequence of forced unnatural human behavior in society, which was mentioned above. This can be presented as follows: by watching a group of people it is possible to see the same physical parameters and one or more masks on their faces. The result is that there is no information neither about the inner essence, true qualities and intentions, nor their real functioning in the environment. However, despite this, one person seriously differs from another, and sometimes cardinally in that he has his own individual psychophysiological structure. If we use the language of biology, then, considering humanity as one biological type—individual archetypal patterns divide people into subtypes. At the same time, by starting to consider people as representatives of different subtype structures, many things become more clear and explicable at once. Including the important thing: another person has his own individual psychophysiological algorithm of work of his psyche and physiology, even if he or she is your child, your parents or another close relative. As a general rule, a person is usually led away from such consideration due to external similarity or blood kinship. Lack of understanding of one's own children, parents or relatives, not to mention friends or acquaintances, still remains. It is necessary to note that unlike representatives of fauna and flora, where reproductive processes have a firm algorithm—conformable gets born, for a human this process is similar only externally, but internally each newborn's archetypal pattern can be completely different, not identical to his or her parents'. If we draw an analogy with the animal world, then a cat can give birth to a hippopotamus, an eagle to a rabbit, an elephant to a mouse. In this case, it seems logical to us to name groups of people, as carriers of characteristic regularities—subtype (subspecies) groups because the term 'subspecies' is a taxonomic unit, which represents totality of separate populations of the same species, representatives of which differ from representatives of other populations of the same species by specific and rather stable characteristics. At least, such consideration of the human community allows answering a lot more questions about principles of interaction in it. In addition, it really becomes possible to begin to consider life of a human (as a representative of a separate subtype) in terms of natural selection in the environment. In this case, a person's life in society becomes presented as it is, without embellishments; in other words—without masks, role-playing games, etc. Moreover, despite many centuries of embellishing it like a Christmas tree, with "bows" composed of masks and "universal human values," life of an individual is presented and begins to finally be considered from the point of the need to fight for survival. The form and essence of this fight can be known values to the carrier himself and to others, and that will help put the fighting process onto a professional basis. Despite the fact that this does not look pastoral, such perception corresponds to reality, where, unfortunately, there is no professionalism. Although realities of human life in the environment, as well as prevailing social trends of competition, aggression and violence, from our point of view, clearly show that there is an ongoing interspecies (intertype) fight. It is veiled, unspoken, but it goes on throughout the history of humanity. Loving spouses quarrel in the kitchen every day; parents and children pass on the problem of "parents and children" like a guidon from generation to generation; best girlfriends compete with each other for men, and men complete with each other for women and a place under the sun; states and neighbors in an apartment complex fight for territory; politicians and mobsters—for spheres of influence; and all people fight with all people for power, which provides the necessary comfortable psychical segments.

Society begins to be considered not as a community of those who seek a civilized zoo, where in locked cages calmly, without causing damage to each other various cute animals neighbor safely, while social laws, norms of common-life, religious requirements, ideology and so on perform functions of grids and locks. On the contrary, it is possible to agree with the banal statement that the community of people more resembles a jungle, in which a ruthless ongoing fighting occurs. Fighting, in which every person is for himself and the winner is the one who is internally ready to never adhere to any rules other than his own and professionally uses external civilized rules of the zoo. Any individual can reject or accept this process, but it is necessary to admit that such state of things is the natural apparatus of our life in the society. It is not quite that what we call "food chain" in nature together with all the consequences, natural selection, fight for a place under the sun, food, territory, females, children, and so on. This aspiration is from nowhere to nowhere, where the struggle for power is a never-ending process.

It is possible to continue to divide unified laws of nature into laws for us, humans and for our "little brothers." However, in our opinion, it is wiser to try to understand what a particular subtype is and realize, become aware of the true state of things in the human community.

By beginning to consider another person in the same way as one would like that person to treat him, a person often gets an unpredictable reaction from the opponent. The result can be anything: from love and affection to hatred, from the level of "a misunderstanding" to the level of a war with total destruction. As a result, a person disoriented by all sorts of conventionalities (filters) accepted in society actually blocks his own view. A person does not even suppose that the style of communication that is comfortable for him, or is simply accepted in the society might be completely foreign to another person, and moreover can provoke aggression in that person (inexplicable even to his own self) towards the opponent. As a result, instead of mutual understanding there is a conflict. This situation emerges when a person who thinks that he has an open mind in relation to the opponent, in fact, does not know, does not want to know or forgets that the opponent has his own soul together with his own preferences, needs and motivation at the unconscious level, the mechanisms and the contents of which are unknown to him, as well as those around him. In the struggle for understanding, people attempt to find keys to one's soul intuitively, by the method of trial and error, analyzing the opponent's reactions to actions and words, sometimes even for years and decades. Not understanding the real reasons behind a conflict, release of aggression (which, in fact, are merely results of incorrect determination of preferences, goals and objectives of an individual, with whom a person tries to interact), one usually uses common conceptions to explain this: "do not do good and you will not get evil", "I wanted the best, but things turned out as usual." When an individual does understand his own self—he gets rid of many physical and psychical issues, as well as the inferiority complex, as a consequence of becoming aware of his unlikeness and "incompatibility" with the general pattern of common human perceptions. Such person begins to clearly understand that commonality, which with all of its comely appearance performs the role of a Procrustean bed, ruthlessly breaking individual structures of those people who fall on it. On the contrary, a clear understanding that each person belongs to a strictly defined subtype in the human population, in our opinion, provides awareness of one's own individuality in the society. It explains to a person the naturalism of his reactions, thoughts, preferences, which are unlike those of other people, and, most importantly, gives him a purpose—an opportunity to become a Human, to whom power is not the goal in itself.

With understanding of individual archetypal patterns as the basis of psyche given to a human by nature from birth, of division of humanity into subtypes, and that a human is regulated by a system of manipulation modes—the scheme of relations "individual—society" emerges in a new light.

An individual turns to society for satisfaction of his needs: physiological, intellectual, emotional, sexual, and psychological. However, how to satisfy them when in reality he feels resistance to his wishes from others, disparity in opinions, preferences, aspirations? Even a child already understands that in order to get fulfillment of his wishes from parents he must influence them in one way or another. And, there are many approaches, ways and methodologies to achieve this. All of them are explained in modern psychology fully enough. In total, all these tricks done by children and adults, which they produce in order to satisfy their needs in society can be described by only one word—manipulation. However, in modern society, among not only so-called ordinary people, and among well-educated, open-minded and creative people as well—complete lack of understanding of the question of what manipulation in human relations is can clearly be traced. Manipulation as such (let alone its true meaning and purpose) is absolutely not understood by most people, and is considered as an unequivocally negative phenomenon. However, such declaratively negative opinion does not stop honest and decent citizens from manipulating one another from morning until night—they just give this process a different name. This creates a distinct impression that some invisible social taboo is placed even on discussion of manipulation as an ordinary, natural and everyday action of people; however, a taboo not in the sense of actions that are being carried out, but in the sense of silence in regard to this issue. So, what is manipulation? Neither the Penguin Dictionary of Psychology, nor the soviet Philosophical Encyclopedic Dictionary provide a definition on this. In the end, our question was answered by literature on psychology of influence. However, even in its titles and prefaces it did not forget to warn its readers that it is necessary to defend oneself from any kind of manipulation. Therefore, at the very beginning of the study of psychology of influence a label has already been attached to manipulation as a phenomenon. Authors of writings about manipulation present it as a hidden management tool of a human against his will, which provides one-sided advantages to a manipulator. For some reason, they delimit manipulation and the concept of "hidden control," which supposedly can have noble goals. They provide various examples as illustrations of the possibility of useful and noble management of another person. For example, a child, gently and wisely guided by a parent with the purpose of education; a subordinate, gently led by his boss in order to improve quality of the production process in the right direction; and so on. However, it is quite alarming how unsure the discussion is carried out in regard to how anyway it is difficult to draw a line between manipulation and so-called "hidden control." For example, who can state with complete confidence that parents do not "softly and gently" cripple a child by their demands? Or is there any guarantee that a "wise leader," supposedly in the name of "better productivity at work" does not pursue his personal goals, in parallel with performance of his official objectives? Therefore, it is possible to conclude that literature, which covers the topic of manipulation, makes a contradictory impression that does not allow distinguishing between "good" hidden management" and "bad" manipulation.

Having concluded that the topic of manipulation in psychology is shamefully hushed up and is insufficiently researched, we decided that our research of this subject from a very different perspective might fill the gap in considerations of manipulations. Especially since consideration of individual archetypal pattern as a human program, which can and must be controllable (manageable and self- manageable) in principle makes the idea of possibility, as well as necessity of manipulations—natural and sensible. Next, we offer to briefly consider what human manipulation modes are from our point of view. Thus, manipulation modes can be both natural and artificial transmittances of properties and qualities of the individual archetypal pattern of a human in a particular scheme in relation to the subject being manipulated in this or that mode. There are three modes to consider and they are defined as follows: "suppression", "balance" and "stimulation." While we briefly describe their properties, in parallel, we will try to show what a person searches for in the society, what it is that he fights for.

Suppression mode is called so because with the use of this mode one person starts to obey another person at an unconscious level and does this with pleasure. "Suppression" occurs when a person onto whom this mode is projected feels as a happily subordinated being. An individual might begin to experience a wide spectrum of feelings towards the carrier of this mode: from respect and love to worship. To some extent this answers the question that has long tormented humanity: "What is love?" On the one hand, people realize that "love" (despite the fact that it appears as a winner among "decorative" objects) is an objectified instinct of reproduction, objectified sexual desire. In reality, all the mysticism and romance of love consists of the following: consciously or not, individual's suppression mode was transmitted towards him, under the influence of which the individual begins to experience a deep feeling of pleasure that hinders attempts to resist subjugation.

Balance mode is the mode of equilibration of human psychophysiology. It represents those qualities and characteristics of the other person, which balance the considered individual, and therefore that person anyway aspires to these qualities. Balance mode is perceived as follows: a person on whom it is projected feels that he is in the most comfortable state, and the person who transmits to him qualities of his balance mode is perceived as a best friend, who understands his aspirations and needs like nobody else. In nature, the following illustrations can serve as examples of such balance: there have been noticed unions between, for example, plants and mushrooms, where both sides supply essential nutrients to each other without mutual detriment. There exist similar beneficial relations among animals, such as the relationship between crocodiles and plover birds, hippos and birds-cleaners, which feed on parasites of the skin of these giants. Something similar can be observed among people, and all this mechanism strongly resembles the description of that what we call by the word "friendship."

During projection on a person of qualities and characteristics of his stimulation mode, he (meaning, the one being impacted) feels irritation, rejection, resentment, and generally total negativity towards the person who projects this mode. In small doses, stimulation mode can effectively and quickly bring a person to specific constructive actions, and in large doses it can cause a very severe outbreak of aggression and rejection, depending on his individual program. Consider an example from the life of animals: next to a pride of lions there are always "grazing" hyenas, which feed on carrion. They constantly provoke lions to leave the prey in order to take advantage of it. This irritates lions, but not enough to make hyenas stop their attempts, while they stay outside of reach. However, when impudent hyenas cross the borders, the end is always the same—a lion annihilates the hyena. A human also cannot be under pressure of his stimulation mode for very long, but if the stimulation process is protracted, the person who transmits this mode automatically gets placed under the category of an enemy, often with the same consequences as in relationships between hyenas and lions. Very likely this example can illustrate more fully why it is proposed to "love your enemy": without reception of his stimulation mode, a human turns into a deedless being, which wants nothing and does not aspire to anything. As we can see, this mode is just as important and necessary for human psychophysiology as other manipulation modes. Without going into details, we will say that with conscious dosing of three positions of manipulation (meaning, by switching between these modes) it is possible to approach questions of management and self-management, as well as questions of building effective communication at all levels and for any purpose. This communication will be effective regardless of social group or which social class its members belong to because these modes work regardless of social status, nationality, religious believes, level of education and intelligence, as they do not pass through the consciousness of a human.

Thus, we tried to consider some aspects of communication flows, what, in our opinion, causes interpersonal conflicts at any level, and what are the basis and causes of various kinds of human relations. Considering apparatus of the society from a somewhat different angle (namely, what is the place of manipulation, which serves as a mean for survival of an individual in the environment), we have made the following conclusions. Manipulation can be considered the main unconscious motivation behind any types of interpersonal relationships. This provides us with a reason to consider interpersonal relations as a system of manipulation. Accepting manipulation as one of the principles of interpersonal relations (that is, considering it as a given and objective reality, which has its own consistent patterns), we think that it is necessary to legalize manipulation principles, mechanisms of manipulative processes as natural and required activity in the society, which, in light of the given research, is a community of manipulators and their subjects.

Bibliography

Yanshina, E. M. (Trans.). (1977). *Katalog Gor I Morej (Shan Hai Tszin)* [Catalog of Mountains and Seas (Shan Hai Jing)]. Moscow: Nauka.

Oshanin, I. M. (Ed.). (1983). *Bol'shoy kitaysko-russkiy slovar' (BKRS)* [Large Chinese-Russian Dictionary (BKRS)] (Vols. 1-4.). Moscow: Nauka.

PART 3

POLITICAL SCIENCE ARTICLES

ARTICLE 1

TERRORISM: A CONCEPT FOR THE ATC (THE COMMONWEALTH OF INDEPENDENT STATES ANTI-TERRORISM CENTER, WWW.CISATC.ORG)

Andrey Davydov
Translated by Kate Bazilevsky.

General Concept

System of threats (hereinafter 'terrorism') is (conditionally) divided into:

- state
- criminal (informal)
- religious (confessional)
- ethnic
- mixed from previous positions in order to unite

Objectives:

- power
- material assets and their equivalents
- ethnic self-sufficiency and domination

Tools:

- fear
- continuously (quickly) improving mechanism of influence in regard to evocation (cultivation) of the first position

Control And Management Of Terrorism (Within The Analytical Department)

The department has two vectors of analytical developments: negative and positive. The first vector (positive) is aimed at development of scenarios in regard to destruction of roots of terrorism in general. The second vector (negative) is aimed at subjugation, control, and management of terrorism according to someone's goals.

The analysis includes:

- development of leaders (at all levels) and environment in the range of their intended use within the framework of the concept stated above
- schematization of mechanisms of splicing terrorism with power and special forces structures
- systematization of information on legal and informal business in order to identify algorithms of "supply" of terrorism
- analysis of export and import of terrorism
- analysis and development of security measures of analytical developments in the relationship "producer—consumer"
- production of a virtual analogue of existing realities with a goal of variative developmental work of manipulations of tactical and strategic impact on terrorism
- development of a system of interaction with other departments
- establishment of branches

Prepared by Andrey Davydov.
Moscow (Russia)
September 2001

ARTICLE 2

INFLUENCE OF PSYCHOPHYSIOLOGICAL SPECIFICS OF A LEADER ON THE STYLE OF POLITICAL DECISION-MAKING

Andrey Davydov

Translated by Kate Bazilevsky.

Introduction

Humanity is accustomed to feeling mature; despite the fact that compared to geomorphological age it is far from it, to put it mildly. Based on this comparison, we can conclude that we are still in childhood state. Maybe to some extent it is due to this that the current level of sensible activity of a human is not free from many natural laws, which often get ignored by theories of knowledge.

Culture undoubtedly includes rules for executions of cogitative operations, interpretations of events, and linking between them. To a great extent these rules determine the worldview inherent in a human. However, this picture has stable central cores, which usually do not change, even despite their apparent inconsistency with requirements of the environment, and these main cores form during very early stages of childhood. One of these cores is a symbol.

Plato, Pythagoras used the term "symbol" and even Homer mentions this word. In his perspective on a symbol, Aristotle ascended to views of his predecessors.

In our own time, we encounter a phenomenon of symbolization, symbolic view and description of the world, as something that is genetically inherent in unconscious structures of the brain. Consequently, a symbol is regarded as a product of symbolic representation of structures of the unconscious.

C. Lévi-Strauss stated that myth structure corresponds to structure of the unconscious (meaning that it reflects the mentality of people who are consciously aware of this myth), and also that cultural phenomena are external expressions of "I," which manifest in the form of symbols and signs.

Philosophy, culturology denotes a series of questions, one of which can be formulated as follows: who is the author of a myth? C. G. Jung has shown that myth creation, as a non-reflexive activity of consciousness relates to the sphere of "collective unconscious." His ideas about heritability of "collective unconscious," the concept of "an archetype" as a great-symbol, as well as the idea that "cultural symbols are important components of our mental apparatus." Another posing of the question was done by E. Cassirer: "If any culture consists of creation of spiritual images of the world, then the goal of philosophy is not to go beyond all these creations, but rather to understand and become consciously aware of the principle of their formation." According to Cassirer, culture is generation of all possible new symbols, and therefore the task of cultural studies, as a philosophical discipline, is to construe these symbols, their essential meaning in order to identify their origin. This origin lies in the general transcendental symbolic function of consciousness itself. Symbolic function of consciousness is manifested as symbolic forms, or (and, it is the same thing) in the form of language, myth, religion, art, and science.

Some answers can be found in C. G. Jung's "depth psychology," namely, in his theory of "collective unconscious" that gave science the concept of "an archetype." Jung states that collective unconscious is identical in all people. This leads to the thought of suprapersonal nature of collective unconscious. However, this does not exclude the thought of individual unconscious and individual archetype, as well as the archetypal complex in the system of collective and individual unconscious. And, that they are that pivot point for an individual, which allows it to have him to have his own individual behavioral style.

Culture Of A Symbol

In the era of classical antiquity, it is possible to highlight three essential points of intuitive, philosophical comprehension of the concept of "a symbol": 1. A symbol is a word, in which properties of things are enclosed in a compressed and hidden form (Democritus); 2. A symbol is a kind of a new beginning, formed from its constituent parts, but not equal to them. The meaning of this principle can only be guessed intuitively (Empedocles); 3. A symbol is that what always carries some mysterious meaning within itself (Pythagoras).

Subsequent development and establishment of the ancient Greek philosophy in the face of its main representatives Plato and Aristotle did not bring anything new and original into the concept of "a symbol." However, symbolic interpretation of Plato's works in the era of Neo-Platonism was the culmination point in history of a symbol. Plato himself interpreted the term "symbol" as a sign ("coin-sign of exchange") or in the form of "Reminiscences of Empedocles" (stories about people-halves, cut in half).

In Aristotle's understanding, a symbol contains a number of meanings, irreducible to each other, but merging into one concept, that is—a symbol. Thus, it is possible to see in a symbol the idea of a treaties and agreement, conjunction of different parts, including those with opposite meanings, some mysterious things that symbolize an idea.

In accordance with teachings of the Stoics about nature-artist, the world seemed to them as created from symbols. Symbols created by a human are connected with divine symbols. The Stoics understood god as nature. And so, Epictetus believed that each symbol is "lovely, splendid," and "significant." With the teachings of the Stoics emerges an interpretation of a symbol as a combination of a phenomenon and its essence because it is possible to judge the inner essence of something only by its external manifestation. Such comprehension of essence is symbolic comprehension. Therefore, in Stoicism, philosophers comprehend the world in symbols. Symbols that come from nature must be perceived by imagination. Soul of a human and its "ruling", "leading" origin has symbols that are rooted in nature. Symbols created by nature are highest, divine symbols, "magnificent in their wholeness." A human comprehends these symbols as conformity of natural phenomena with their essence. Therefore, the task of philosophers Neo-Platonists is to uncover the depth of conceptions of nature behind its external manifestation, which in turn is possible only by thinking symbolically.

Iconic side of a symbol, its simplicity and accessibility have become alien to philosophical aspirations of Neo-Platonists. Not a symbol, but symbolic comprehension of truth—this is what becomes the basis of philosophizing of Neo-Platonists. Plotinus, the founder of this philosophical direction (Neo-Platonism), never mentions the term "symbol," while considering myths in their symbolic aspect. Since a symbol (here associated with a myth) is just a kind of a shell, surface of a myth, but not its essence. Symbols arise from thoughts of a philosopher, which consist of two halves: his own and mythical

Porphyry is another Neo-Platonist, a disciple of Plotinus, and the author of the famous commentary on the work of Homer called On The Cave Of The Nymphs. He believed that it is "through dreams and symbols" that a human gets notified of divine will. According to Porphyry, to express "ancient wisdom" means "to speak figuratively," using "secret (mythological) symbols." If Porphyry's mentor Plotinus has not yet dared to use the word "symbol" even in the sense of a myth—after his student, the term "symbol" quite firmly and widespread becomes a part of the common Neo-Platonic terminology.

Neo-Platonists, Plato's commentators Olympiodorus, Damascius, John Philoponus, as well as Simplicius (a commentator on Aristotle) resort to the concept of "a symbol" in analysis of not only myths, but already philosophical texts as well. For example, Olympiodorus discusses the dog in Plato's "The State" as a "symbol of sensible life." The idea that the practice of closing the eyes (as well as the mouth) of the deceased is a symbol of energy retention also belongs to the same Plato's commentator.

Thus, inevitability of education and use of symbols in their gnoseological significance settles in antiquity for the first time. Damascius also believed that since symbols are "crystal clear" and do not differ from gods, then they are gods. They are given to people as "symbols of the gods" or "unspoken symbols of unspoken gods" as a result of spiritual emanation from gods of intelligible world into the existing world. Thus, Damascius equated a symbol with divine principle.

Simplicius, another famous commentator on the ancient Greek classics, also thought that the wisdom of the ancients was in their symbolism. Thus, commenting on the work of Aristotle called "Categories," Simplicius wrote that the wisdom of Aristotle is in that he was able to reduce the diversity of the world to ten logical categories and collected "all essences in one supreme essence." Ten categories of Aristotle, in the opinion of his commentator, is the symbol of supreme essence, which in its turn is their substance, or "support in the thought process." And here, as it can easily be noticed, implicitly lies the idea of identicalness of a symbol and the divine principle (supreme essence). At the same time, both Damascus and Olympiodorus also adhere to tradition that stemmed from Aristotle and Democritus, and believe that a name is a symbol. And, a name is symbolic only if it is identical to the essence of the carrier. Since a name contains the essence of a phenomenon, then knowledge of the world is knowledge of the name–symbols. Knowability of the world that is infinite in its manifestations is carried out by means of categories set by Aristotle because supreme essence is symbol and substance of these categories.

It is possible to draw a conclusion: for the first time gnoseological value of a symbol is pointed out in the philosophy of Simplicius; namely–the possibility to reduce infinity of world phenomena to logical categories by means of symbols.

Spiritual quests of latter Neo-Platonists were completed in the philosophy of Proclus. That is how Proclus, the concluder of Neo-Platonism (a teaching, the central category of which is a symbol), provided a conclusive definition of this concept. His thoughts on nature and essence of a symbol are described in the commentary to Plato's "The State." Appearance and presence of symbols in the human world that is being comprehended and created, Proclus explains by basic human needs to talk about god, the divine. However, this cannot be done in representations of human life, but is possible by the means of symbols. It is through symbols that divine potency reveals itself and without a hinder gets transmitted to a person in the form of divine spirit. According to Proclus, even the gods themselves rejoice hearkening these symbols. Following Plato, Proclus also states that symbol cannot be imitations of that what they are symbols of. This thought actually embodies the aesthetic concept of Neo-Platonists. A symbol is contraposed to an image because the latter is the result of imitation. An image must be similar to an exemplar, and a symbol—identical to "the truth of being," which can be achieved not by copying the object, but, on the contrary, by finding an opposite origin in it. Combination of opposite beginnings is achieved only in a symbol. Therefore, the truth in a symbol is revealed gradually, through combination of opposite beginnings. This makes a human, his idea break through the "dark," go from incomprehensible to "transparency," to simplicity of the truth, in which divine potency is enclosed, emanated.

Thus, based on the example given above, we can see how slowly, as if swaying between literalism and conventionalism, the concept of "a symbol" takes its shape; its formation was completed in Neo-Platonism and in this form—passed into new Europe, and got fixed in the European tradition.

I. Kant, the founder of classical German philosophy, interprets a symbol as a special kind of representation, through which gets transmitted something that cannot be summarized by the means of discursive sense. This notion of "a symbol" comes from Kant's separation of two kinds of knowledge: sense and reason. Reason, real reason, is intuitive sense, meaning—non-discursive. According to Kant, the discursive sense is presentation of a whole as being dependent on parts, while reason (intuitive sense) is comprehension of things as they are themselves, meaning pre-imaged, where parts (their properties and relationships) are conceived as being dependent on the whole. Reason comes from synthetic general (contemplation of the whole) to specific. Therefore, according to Kant, a symbol is an "idea of reason" given in contemplation. Such contemplation is "when under the notion that can be conceived only by reason, and to which no sensual contemplation can correspond, no contemplation can be brought."

Thus, according to Kant, a symbol is a sensual way of presenting ideas of reason, while comprehension of a symbol is intuitive comprehension of an idea, which is infinite and inexhaustible in its ineffability because it is impossible to effable.

A symbol is the central category of Schelling' philosophical system. Following Schlegel, he believes that substantial basis of a symbol is the idea of infinity, which Schelling also interpreted as absolute identity of the ideal and the real. The main particularity of all that what is perceived as a symbol is its inconceivability, its essential and indissoluble bond and unity with the idea. At the same time, an idea in a symbol is embodied in an image, and Goethe put it very well: "Symbolism transforms a phenomenon into an idea, the idea into an image in such a way that the idea stays in the image as infinitely operational and incomprehensible. Even expressed in all languages, it still remained ineffable."

According to Schelling, the most complete and in-depth reproduction of the essence of the world is realized in art, to which a symbol is a fundamental concept. However, since historically art came out of mythology, then mythological symbol is the basis—meaning, a type of a prototype for a symbol in art. Moving on, Schelling states that any true art must be completed with creation of its own mythology. Thus, Schelling writes: "Every great poet is intended to turn a whole part of the world revealed to him into something and to create a mythology from the material."

Therefore, in order for an image to become a symbol, it is necessary to identify its symbolic nature from the context. It is the context in which a sign functions that serves as a link between an image and a symbol. Therefore, it is possible to consider Hegel's statement (that "a symbol in general ... can be regarded as a sign or a mark, in which the relationship between the meaning and its expression is quite arbitrary") to not be entirely correct.

To Hegel, a symbol is a sign that by its external form expresses a certain meaning, which we understand in the presence of a particular context (culture) as symbolic content of this sign. Hegel wrote that a symbol is "some contemplation, own definiteness, which in essence and concept is more or less that content, which it, as a symbol, expresses; on the contrary, when it comes to a sign as such, then the own content of contemplation and that of which it is a sign of—have nothing in common with one another."

Based on Hegel's understanding of philosophy as "knowledge in concepts," it is possible to assume that as such phenomena must appear the most general concepts, developed in culture being studied, meaning—its categories. "Understanding of cultural phenomena in comprehension implies consideration of mediated human senses and leads to introduction to the theory of such categories as "ideal", "tradition", "meaning of existence", "value", and so on." However, the term "introduction to the theory of categories" does not correspond to reality, that is—consideration of their ontological status. "Categories are not introduced into culture from the outside, from a philosophical or scientific reflection, but are produced within it." They act in culture at the unconscious level or, using Hegel's expression—"instinctively." Therefore, supposition regarding categories of cultural phenomena must be transferred to the category of a statement: "categories developed by culture are that form, by which comprehension of culture gets attained." It is necessary to emphasize once again that this "comprehension of culture" is a characteristic only of philosophical consciousness, and that is no accident.

It is known from the history of philosophical thought, philosophical consciousness turned to the problem of categories of culture relatively recently. This is explained by the circumstance that the very notion of culture becomes an object of philosophical reflection only in the New times.

As it is known, there are over 250 definitions of culture in literature. Such a multitude of definitions is caused by that the concept of "culture" has become the subject of analysis of not only philosophy, but a number of other humanitarian disciplines as well.

E. Cassirer is the author of three-volumes of Philosophy Of Symbolic Forms, which earned him a "lifetime glory as a philosophical classic." This work, along with the work of O. Spengler called Decline Of The West, turned out to be large-scale claims for comprehension of the problem of categories of culture. According to Cassirer's opinion, in order to disclose the essence of human existence, it is necessary to first of all turn to research of culture that "accumulates all its universal properties." Therefore, "the main question to be resolved at the start of philosophical analysis of culture is the question about the possibility of discovering one common thing that lies at its base, behind all its diverse manifestations." For this, Neo-Kantian Cassirer modified Kantian's "priori form" (meaning a formal synthesis of sensual multiformity) and began to interpret it as a "symbol," giving it the status of a central category of his philosophizing. Cassirer interprets human's symbol-forming ability as "a priori property of a human." Those diverse spheres of cultural expression (language, religion, myth, art, science) that Cassirer calls "symbolic forms" cannot be reduced to a common. According to Cassirer, each "symbolic form" (a special symbolic function of consciousness) is realized in a special type of reality—language, myth, religion, art, science. This allowed him to come to the idea that a human is not animal racionle, but first and foremost is animal symbolium.

For philosophical analysis of culture, a theoretical synthesis of categories of culture is important. One of the first to formulate the problem in this way was the largest representative of philosophy of life—W. Dilthey. To Dilthey, philosophical foundation of culture is "hermeneutics," which he understands as "a mean of recreation of unique and self-enclosed cultural worlds of the past." The philosopher explains this (hermeneutic) understanding of culture by that the subject of culture contains human significance, meaning: "he already appears as a certain result of some conscious synthesis, categorization of the world on the part of another subject." This determines specificity and complexity of analysis of phenomena of culture as opposed to natural phenomena. According to Dilthey, a central role in humanitarian cognition plays the category of "significance," on which, as suggested by the philosopher, all others depend.

P. Ricœur, leading representative of the French school of hermeneutics, notes that language is a secondary understanding of reality, which, however, initially has a symbolic function. The logic of Ricœur's statements seems quite convincing, and therefore, "if understanding is a way to comprehend individual phenomena of existence, then it is necessary to go "behind" linguistic meanings into a richer world of symbolism in general." The internal logic of formulation of the problem of understanding of cultural phenomena forces Ricœur to turn to analysis of the concept of a symbol: "...I give the word 'symbol' a narrower meaning than writers who, like Cassirer, call any perception of reality by means of signs symbolic, from perception, myth, art to science, but a much broader meaning than those authors who stem from Latin rhetoric or Neo-Platonic tradition. I call a symbol any structure of value, where direct, primary, literal meaning at the same time has another, indirect, secondary, allusive meaning, which can be understood only through the first. This range of expressions with double meaning makes up the hermeneutical field." As a result of cultural studies, Rickert concludes that "a symbol leads to the thinking process."

Thus, structuralistic interpretation of "a symbol," as a product of unconscious activity or non-reflexive consciousness, becomes the object of reflexive consciousness in hermeneutics. This means that unconscious activity of primitive people embodied in their myths can and should become the subject of analysis—identification and interpretation of meanings of cultural symbols. Ontological basis of this thesis is provided by hermeneutics of a symbol since "The statement of hermeneutics as a philosophical discipline is based on that in nature of reflective thinking it is possible to find principles of logic of double meaning—logic of interpretation of symbols. However, it is not formal logic, which deals exclusively with direct and exact definitions and mathematical expressions. This logic is transcendental: it uncovers not only categories of scientific comprehension of nature, as Kant tried to execute, but also the range of possible arrangements of human existence. All concrete richness of our conscious existence—"I am"—are "stored" in a hidden "oneiric" form in symbolic reserves of human culture."

"Personal Unconscious"

An attempt to combine the deep meaning of symbols with their external features can be described as consideration of a symbol as a product of symbolic representation of structures of the unconscious. G. Krauss and E. von Hartmann developed the "unconscious" as a set of psychical processes that are not represented in the consciousness of a subject and as a universal base of existence. Development of this trend continues in C. G. Jung's deep psychology. In his discussion, Jung bases on the fact that "certain ideas exist everywhere, at all times" and "they are not created by individuals, but come into, even forcefully invade the consciousness of an individual." Thus, Jung as if brings empirical, psychological justification under the philosophy of Plato. Jung calls these "non-creatable ideas"—"the collective unconscious" and states that it is identical in all people. According to Jung, it turns out that nature of the "collective unconscious" is superpersonal. Jung's main achievement is in that he uncovered a "collective" character of a wider concept of "unconscious" than those already mentioned by G. Krauss and E. von Hartmann.

In order to be able to discuss content of the collective unconscious, Jung introduced the concept of "an archetype," which he characterized as an explanatory description of Plato's "eidos." However, Jung does not provide an exact definition of the concept of "an archetype." Possibly this is explained by the general methodological approach of the creator of "depth psychology." Since the term "archetype" reflects the content of the "collective unconscious," which in turn manifests as a symbol, therefore, content and definition of the concept of "an archetype" are symbolic. Thus, Jung describes the term "archetype" as a "primeval image," a tendency to form certain perceptions, instinctive vector, thought-form, particular impelling energy, unity of image and emotion, etc.

In its original meaning, the "unconscious" was understood as a kind of receptacle, where everything that has been suppressed or forgotten collects, that is—perceptions ousted from consciousness. S. Freud, the founder of psychoanalysis, also shared such understanding of nature of the unconscious. However, he considered "libido," meaning—sexual desire, ousted into the unconscious as the central concept. Since the noted perception of the unconscious has a personal nature, Jung called it "personal unconscious" in contrast to "collective unconscious." According to Jung, "personal unconscious" lays on the surface of the unconscious, as it is acquired from personal experience. "Collective unconscious" is something innate, so it represents the deep layer, the very essence of the unconscious.

Freud noted manifestation of the "collective unconscious" as "dream images." However, he called them "archaic remnants," meaning—certain islands of psyche, which somehow survived in human brain since the most ancient (primeval) times. However, Jung's clinical studies have shown that these images are "not in any way lifeless or meaningless remnants," but "are an integral part of the unconscious." Such statement became possible only after Jung established that dreams "have a defined, clearly purposeful structure that indicates its latent idea or intention," meaning that a dream also has nature of a symbol. Based on this understanding of structure and function of a dream, Jung sets a task "to research the ability of a human to produce symbols."

A human is unable to consciously control the brain's dreaming activity, as this activity more belongs to the sphere of the unconscious. At the same time, not only dreams belong to the sphere of the unconscious, but also all non-reflexive activity of consciousness and, first of all, myth-making. What indicates that we are dealing with the unconscious? Jung's response to this is as follows: "Whatever the unconscious may be, it is a natural phenomenon that reproduces meaningful symbols." Based on this thesis, it can be argued that a symbol is a product of symbolic representation (structure) of the unconscious.

It is known that along with individual symbols there are also collective ones, on which myths and religion are based. Regardless of whether a symbol is individual or collective—it arises from the unconscious, as, according to Jung, it is impossible to consciously invent a symbol. A symbol is always greater than the rational concept because it is rooted in the unconscious, in the psyche. As it is known, psyche is more than consciousness, as it not only gets acquired through life experience, but it is also innate. A human is not born with "empty psyche," while consciousness can be compared to a "clean slate." Psyche is the result of evolution of all humanity, races, nations, tribes, clans, families, and animal life. A symbol is rooted in the unconscious. Jung writes: "Just as a plant produces flowers, psychical raises its symbols," and "cultural symbols are important components of our mental apparatus."

Since a multitude of archetypes exists, then the most important unifying factor is their (archetypes') informational-emotional-energetic characteristic. This is related to that a "word" itself can be considered as a carrier of information; it has a certain psychical energy, which in turn is supported by the fact that for people who live in a primitive state (society)—things of the surrounding world are still endowed with psychical qualities. Thus, the term "archetype" can be understood as an integral part of the collective unconscious, having informational, emotional, psychical, and energetic effect on consciousness. Since an archetype is also the content of collective unconscious, then the most typical of them can be identified from this content. It should be noted right away that this refers to the most typical archetypes. Since the term "archetype" is itself symbolic, this makes its content barely detectable. Therefore, when referring to an archetype, we are talking about a symbol, and the only difference is that a symbol is a "polished" archetype, meaning—an archetype that got a specific form. For example, the idea or the image of Abraxas is an archetype that, like every genuine archetype, has enormous psychical energy. This energy, going from the collective unconscious to conscious (meaning—affecting it) makes people, whole nations create specific images of gods. Almost every nation has created its God. Many gods, each of which is a certain symbol have been created from one archetype. God is a psychical phenomenon, and not intelligible. Thus, this is how an archetype must be understood—as a psychical phenomenon that represents part of the "collective unconscious." An archetype is a bundle of certain kinds of symbols. In other words, a countless multitude of symbols can be established from one archetype. However, all these characters will rise and unite in an archetype that is common for them, part of the archetypal complex.

Since previously we defined an archetype as a property of the unconscious that has informational, emotional, psychical, and energetic effect on consciousness, meaning as a kind of psychical energy, then, consequently, in its essence an archetype, as a form of energy, is neutral in respect to ethical and aesthetic categories. And, this means that such sphere of psychical activity as, for example, artistic expression, probably cannot be considered within the framework of analytical psychology? The answer to this question can be found in a symbol because true art and deep psychology intersect in it. Since a true work of art is symbolic, then analytical psychology allows better understanding of cause of appearance of such work of art. In analytical psychology, "the process of artistic image-creation," not aesthetic value of art should be considered. Jung writes: "Unborn work in the painter's soul is an elemental force that makes its way either tyrannically and violently, or with that inimitable cunning, with which nature is able to achieve its objectives, not caring about personal wellbeing or grief of a person—carrier of creative beginning."

Future work in the soul of a painter, in his unconscious, is commonly called an autonomous complex. This complex is autonomous; it is completely self-dependent and has a certain energy level. A finished work of art is a consequence of realization of the autonomous complex. At the same time, two completely different ways of artistic image-creation are possible. In the language of analytical psychology, they are commonly called introverted and extroverted. Jung introduced both terms. Thus, he writes: "Introverted way is characterized by assertion of a subject with his conscious intentions and goals, as opposed to claims of an object; extraverted way, on the contrary, is marked by the subject's submission to requirements of an object." At the same time, Jung denied participation of a painter's personality in his work of art, comparing painter's personality with soil in which grows a plant—a work of art. As a plant grows from a seed, a work of art stems from an individual, "personal" archetype." Authenticity of a work of art consists in that if it goes beyond the personal, then it is something superpersonal. Such work differs by "strange" and unexpected form, which at the same time speaks in a multitude of languages understandable by us. This work of art is a true symbol. Symbolic work, in contrast with non-symbolic, "excites us more", "gimlets deeper." Genuine artistic symbol is based on an archetype, and the latter is "psychical residue" of innumerable experiences of similar type. Artistic symbol is realization of that possibility, which we "from the most ancient times inherited in a certain form of mnemonic images or, to put it in anatomical terms, in the brain structure."

Art has great social significance, as it wholly and intuitive captures that what is happening in the collective unconscious. Thus, we can say that any creation—mythical, religious or artistic—is representation of "structures of the brain."

Structuralism, as a philosophical direction, emerged in the 1960s in France and owes a lot to the French sociological school. As C. Lévi-Strauss himself notes, the main role in formation of his concept was played by the French sociologist Émile Durkheim, who's merit was in that he made a sharp distinction between individual and collective consciousness. In the philosophical aspect, structuralism is based on three directions that have emerged in the philosophical thought of the XX century.

The first is symbolic philosophy of E. Cassirer; the second is analytical philosophy of C. G. Jung; and the third is structural linguistics of F. de Saussure.

From analytical psychology, structuralists borrowed the idea of the unconscious in the form that was understood by Jung, meaning in the form of "collective unconscious." Just like Jung, C. Lévi-Strauss considered mental structures. However, Lévi-Strauss was against Jung's idea about constancy of archetypes and hereditary mechanism of their transmission. In structuralism, permanent inherited is not the content of the "collective unconscious" (meaning, an archetype), but "the structure of the unconscious." Structuralists borrowed the idea of such a structure from structural linguistics. C. Lévi-Strauss writes: "Unconscious mental activity consists in giving form to content," and, in his opinion, this "brilliantly opens up with research of symbolic functions in that form, in which it gets expressed in a language."

As it is easy to see based on statements above, there are as if two kinds of structures: structure of the human reason (mind) and structure of culture created by a human. In this case, the concept of "culture" includes "science," and thus—comprehension of physical reality, nature, which also has a structure. A particularity of structuralism is that an equal sign is set between these two structures, and then it is stated that there is only one unconscious structure of the mind—structure of the unconscious—that is not consciously perceived by a human, but is projected by him onto the surrounding empirical reality, the result of which is emergence of personal cultural phenomena. In order to obtain the structure of the unconscious, structuralism proposes to consider cultural phenomena, structure of which, in their opinion, is the structure of the unconscious. Thus, the philosopher Lévi-Strauss tries to do this by the way of studying myths; psychoanalyst Jacques Lacan—by the way of studying psychology of the unconscious; cultural historian Michel Foucault—by studying madness and crime; literary critic Roland Barthes—by studying artistic literature texts.

The structure of the unconscious is a kind of a "tool," by the use of which all information that comes from the outside gets structured, obeys the law of structure. Since the structure of the unconscious remains constant, while the content of structured information changes, then, consequently, structure performs a symbolic function. Therefore, at its base, a myth has a symbolic nature. It equally explains past, present and future. In this regard, a carrier of the structure of the unconscious—an individual—becomes interesting.

Pythagorean Approach In The Theory Of Personality

Following the practice of Pythagoreans, terms "a human, "an individual", "a leader" (including political) can be regarded as equivalent, and, going forward, the word "individual" can be used. Unlike Pythagoras, this will allow to not extend the esoteric symbolism, but to consider an "archetype" as a "great-symbol" in chimeric embodiment, based on which an individual intuitively approbates his own individual archetypal complex within the social apparatus in order to cognize (know) his own self.

Intuitive approbation of individual archetypal complex has a random effect on processes of any environment. In this case, it is possible to observe significant dependence of an individual on the environment, which enables psychologists, sociologists, and others to conclude that an individual is often defined as a human in the totality of his social, acquired qualities. Consequently, these numbers do not include those particularities of a human, which are genotypic or physiological and do not depend on life in the society. In many definitions emphasis is placed on that these numbers do not include psychological qualities of a human, which characterize his cognitive processes or individual style of activity, except for those that manifest in regard to people, in the society. The concept of "an individual" typically includes properties that are more or less stable and testify to individuality of a person, defining his actions that are significant to people. The conclusion is that an individual is a human taken in the system of his psychological characteristics, which are socially determined, manifest through connections and relations that are social by nature, are stable, and determine moral actions of a person, which have a significant meaning to himself and others.

However, if we assume that a human is formed not only by the environment, and individual archetypal complex largely, if not primarily, dictates psychophysiological algorithms of a person, but depending on the extent that is necessary for survival in the environment, then this assumption grows into certainty. In this case, it is possible to offer a definition of "an individual" that is somewhat different from the previous: knowledge of the individual archetypal complex, its conscious approbation in society leads to organic merging of a symbol and a human and multivariance of conscious corporate (synergistic) transformation of the environment. And, representation of personal archetypes in the environment can be understood as an attempt to process subjective images into objective reality. Form varies: symbols, hieroglyphs, conventional signs of this reality. In regard to construction of the world within and outside of ourselves will be said below.

Neurophysiologists Breyn, Adrian, and others assimilate the relation of neurodynamic structures of the brain to external stimuli (phenomena of the external world) with the relation of a geographical map to a particular territory. Therefore, they termed certain state and distribution of nerve impulses through the cortex as "cortical map," and the theory itself got the name of "the theory of cartography." Thus, a representative of this direction, neurophysiologist Adrian writes: "As for the brain, the function of sense organs or receptors is to construct it in a map of certain physical events, which take place on the surface of the body in order to show what is happening in the world outside of us." In other words, the function of the cerebral cortex is to perform symbolic representation of the external world, and thus generate symbolic information about it. This once again confirms the idea that knowledge of individual archetypal complex, its conscious approbation in society leads to organic merging of a symbol and a human and multivariance of conscious corporate (synergistic) transformation of the environment.

Conclusions

Through understanding of what a "symbol" is, of its history, cultural application—it is possible to conclude that the influence of psychophysiological specifics of an individual on the style of decision-making mainly depends on the interaction of his individual archetypal complex with the environment. And, this requires efforts of philosophical, culturological, psychological, and other scientific directions in a single, synergistic approach to the problem of individuality.

German physicist Hermann Haken proposed the term "synergy" to describe the aggregate, collective effect of interaction of a large number of subsystems, which lead to formation of stable structures and self-organization in complex systems. Originating from non-equilibrium thermodynamics, synthesis of knowledge of natural sciences, synergy orients on the disclosure of mechanisms of self-organization of complex systems—natural and social, as well as those created by human hands. Together with synergy came the understanding of unity of inorganic and organic world, understanding that alternation of chaos and order is a universal principle of world order. According to academician N. Moiseyev: "everything we observe, all that we are involved in today are only fragments of a single synergistic process..."

The perspective that society is a social machine that operates according to "objective laws" is a pre-synergistic perspective. Modern natural sciences, science in general, and social life force to learn new synergistic tools of thought. Most likely, the approach to the problem of influence of an individual on the environment through understanding that this influence depends mostly on personal parameters, decryption of which requires joint efforts of various scientific directions—the approach is synergistic. This might allow to more clearly indicate the place of a modern human in the emerging informational society and to identify correlates of an individual and the social environment.

Bibliography

Windelband, B. (1910). *Filosofiya kul'tury i transtsendentalnyy idealizm* [Philosophy Of Culture And Transcendental Idealism.]. Moscow: Tip. "Pechatnoye Delo."

Rickert, H. (1911). *Nauki o prirode i nauki o kulture* [Natural Sciences And Cultural Sciences.]. St. Petersburg: Obrazovaniye.

Goethe, J. W. (1964). *Izbrannyye filosofskiye proizvedeniya* [Selected Philosophical Works.]. Moscow: Nauka.

Kant, I. (1965). *Soch. v 6-ti t.* [Comp. in VI Vols.]. Moscow: Mysl.

Hegel, G. W. F. (1969). *Estetika. Soch. v 4-kh t.* [Aesthetics. Comp. in IV Vols.]. Moscow: Iskusstvo.

Kravchenko, A. (1970). *Kassirer ob iskusstve kak simvolicheskoy forme* [Cassirer On Art As A Symbolic Form.]. Moscow: Nauka.

Abramova, A. Z. (1972). *Drevneyshiye formy izobrazitelnogo tvorchestva* [The Most Ancient Forms Of Visual Art.]. Moscow: Iskusstvo.

Averintsev, S. S. (1972). *Analiticheskaya psikhologiya K.G.Yunga i zakonomernosti tvorcheskoy fantazii* [C. G. Jung's Analytical Psychology And Patterns Of Creative Imagination.]. Moscow: Iskusstvo.

Stolyar, A. D. (1972). *O genezise izobrazitelnoy deyatelnosti i yeye roli v stanovlenii soznaniya* [On Genesis Of Visual Activity And Its Role In Development Of Consciousness.]. Moscow: Iskusstvo.

Losev, A. F. (1976). *Problema simvola i realisticheskoye iskusstvo* [The Problem Of The Symbol And Realist Art.]. Moscow: Iskusstvo.

Tahoe-Godi, A. A. (1980). *Termin "simvol" v drevnegrecheskoy literature* [The Term "Symbol" In The Ancient Greek Literature.]. Moscow: Izd-vo MGU.

Vasileva, T. Y. (1980). *Problema germenevticheskogo metoda v sovremennoy burzhuaznoy filosofii* [The Problem Of Hermeneutic Method In Modern Bourgeois Philosophy.]. Filosofskiye nauki.

Bulatov, M. A. (1983). *Kategorii filosofii i kategorii kultury* [Categories Of Philosophy And Categories Of Culture.] Kiev: Nauk. dumka.

Filosofskiy entsiklopedicheskiy slovar [Philosophical Encyclopedic Dictionary.]. (1983). Moscow: Sovetskaya entsiklopediya.

Ivanov, V. V. (1983). *K. Levi-Stross i strukturnaya teoriya etnografii* [C. Lévi-Strauss And Structural Theory Of Ethnography.]. Moscow: Raduga.

Lévi-Strauss, C. (1983). *Strukturnaya antropologiya* [Structural Anthropology.]. Moscow: Nauka.

Meletinsky, E. M. (1983). *Mifologiya i folklor v trudakh K. Levi-Strossa* [Mythology And Folklore In The Works Of C. Lévi-Strauss.]. Moscow: Nauka.

Turner, B. (1983). *Simvol i ritual* [Symbol And Ritual.]. Moscow: Nauka.

Aristotle (1984). A. I. Dovatur (Ed.). *Soch. v 4-kh t.* [Comp. in IV vols.]. Moscow: Mysl.

Bell, D. (1986). *Sotsialnyye ramki informatsionnogo obshchestva* [Social Framework Of The Information Society.]. Moscow: Progress.

Novikov, M. P. (Ed.). (1986). *Ateisticheskiy slovar* [Atheistic Dictionary.]. Moscow: Politizdat.

Filosofskiy slovar [Philosophical Dictionary.]. (1987). Moscow: Politizdat.

Gadamer, H. G. (1988). *Istina i metod: Osnovy filosofskoy germenevtiki* [Truth And Method: The Basics Of Philosophical Hermeneutics.]. Moscow: Progress.

Batkin, L. M. (1989). *Italyanskoye Vozrozhdeniye v poiskakh individual'nosti* [Italian Revival In Search Of Individuality.]. Moscow: Nauka.

Freud, S. (1989). *Psikhologiya bessoznatelnogo* [Psychology Of The Unconscious.]. Moscow: Prosveshcheniye.

Svasyan, K. L. (1989). *Filosofiya simvolicheskikh form E. Kassirera* [E. Cassirer's Philosophy Of Symbolic Forms.]. Yerevan: Izdatelstvo AN Armyanskoy SSR.

Losev, A. F. (1991). *Logika simvola* [The Logic Of The Symbol.]. Moscow: Polit. lit.

Losev, A. F., Meletinskii, E. M., Riftin, B. L., Toporov, V. N., Ivanov, V. V., Averintsev, S. S., ... Afanasyeva, V. K. (1991). S. A. Tokarev (Ed.). *Mify narodov mira* [Myths of the World] (Vols. 1-2.). Moscow: Sovetskaya Entsiklopediya.

Rakitov, A. I. (1991). *Filosofiya kompyuternoy revolyutsii* [Philosophy Of The Computer Revolution.]. Moscow: Politizdat.

Rubtsov, N. (1991). *Simvol v iskusstve i zhizni* [Symbol In Art And Life.]. Moscow: Nauka.

Sovremennaya zapadnaya filosofiya: slovar. [Modern Western Philosophy. Dictionary.]. (1991). Moscow, Politizdat.

Hall, M. P. (1993). *Entsiklopedicheskoye izlozheniye masonskoy, germeticheskoy, kabalisticheskoy i rozenkreytserovskoy simvolicheskoy filosofii* [Encyclopedic Outline Of Masonic, Hermetic, Qabbalistic, and Rosicrucian Symbolical Philosophy.]. Novosibirsk: VO "Nauka", Sibirskaya izdatel'skaya firma.

Heidegger, M. (1993). *Raboty i razmyshleniya raznykh let* [Works And Thoughts Of Different Years.]. Moscow: Gnozis.

Kerlot, K. E. (1994). *Slovar simvolov* [Dictionary Of Symbols.]. Moscow: REFL-book. ISBN: 5879830144

ARTICLE 3

ESSENCE OF POLITICAL IDEOLOGIES AND THEIR ROLE IN THE HISTORICAL PROCESS (POLITICAL HISTORY OF RUSSIA)

Andrey Davydov

Translated by Kate Bazilevsky.

Introduction

If we rely on logic of history of political and legal teachings, then this implies that this history began in Rus' at the time when pagans were huddled into a river. This fact in itself is of little interest. What is of interest is the following: what is the essence of this fact?

In his time, T. Mann expressed a thought that if humanity once again were put in pre-civilized conditions—the first thing it would do is create mythology. Mythology, as it is known, operates on images. In terms of survivability, there is nothing comparable to the degree of survival that an image has. This can be traced back through heraldry of many countries, including Russia. At the same time, as strange as this might seem at first glance, military authorities of all countries in the world today can be nominated for the palm of victory in the use of mythological images.

An image has a very large informational capacity and, at the same time, it is multifunctional, very convenient, and aimed directly at psyche of a person, who under the influence of imagery (mythology) carries out his physical actions only within the limits of the received image. Naturally, an image can be intended not only for an individual or a group, but also for social, ethnic influence. In the history of political and legal doctrines of Rus' it is interesting to trace the work of the system of images from the point of view of justification for existence of monarchy, relations between monarchy and Christianity, monarchy and philosophy, ideology and politics, real and virtual worlds.

Politics, according to statements made by political scientists, as well as directly or indirectly related people, claim relation to art. If this is so, maybe it is worthwhile to consider politics alongside creative works of people of literature, poetry, painting?

Paganism And Christianity On Rus' Territory

According to one of encyclopedic formulations, an image is the result and the ideal form of reflection of objects and phenomena of the material world in human consciousness. An image, an artistic method and form of understanding of reality in art are characterized by indivisible unity of feelings and meaningful aspects.

Domination of pagan image (system) was full and unconditional until the emergence of state of Kievan Rus' in 882 and adoption of Christianity in 988. Tendency to consider the phenomena of the physical world as products of creativity made by some creatures who have a will or by people in order to achieve specific human goals—this was habitual and expedient.

Domination of mythological (pagan) images allowed gradual individual development to be identic to development of the whole biological type. The main particularity of the mythological pagan system was that people perceived its positions and information as indisputable truths; truths that allow a person to have a whole, non-contradictory worldview.

A myth is characterized by astonishing stability. This stability has several explanations. Firstly, the ancient people had inherent eidetic or holistic memory, which allows very precise memorization of texts and thus an ability to reproduce them, and that is quite important when orally retelling mythological stories. Secondly, exact reproduction of myths can be explained by their topics—plots of myths of all time are very similar to each other. Thirdly, stability of myth can also be explained by specifics of social control in the ancient society: it mainly supported and encouraged not creative, but reproductive actions of people. However, during the period of transformation of Rus', pagan orientations became no longer necessary in the future state structure.

The pagan system is a fairly strict system of mythological images, but this did not prevent individual freedom of interpretation by users, which in turn contributed to optimization of connections between its elements. In total, all these positions hindered the formation of statehood in Rus'. Total replacement of the primary mythological system with a new, more structured one can serve as a way out. Christian religion became this system, and first of all it began to change or destroy practically all of the ancient mythological order. Under the conditions of formation of feudal society and emergence of statehood complicated by external danger—this allowed to continue to form and develop the ancient Russian political and legal idea of independence of Russian state, its unity, and strong princely power.

The origin of the ancient Russian political ideology is connected with annals that emerged in the first half of the XI century. Based on orally transmitted stories, annalists (mainly of Kiev-Pechersk Monastery) tried to restore the historical past and explain the present Rus'.

Christianity And Monarchy

In the middle of the XI century appeared the first purely political work called Sermon On Law And Grace by Metropolitan Of Kiev Hilarion, who sought to theoretically justify independence of Kiev state from Byzantium as well as the idea of strong princely power. Sermon On Law And Grace presents a fantastic concept of world history written in a religious tone and divided into two periods: the Old Testament and the New Testament. The period of the Old Testament is a period of God's chosen Jewish people, a period of subjugation to law. Its successor—the period of the New Testament is a period of grace, when Christianity became the domain of all people, who took it freely and voluntarily. The Russian people have joined Christianity on their own accord and not under the influence of Byzantium. Thus, entered into the equal rights family of nations and did not need anyone's guardianship. Hilarion praises Prince Vladimir Sviatoslavich (Christianizer), the powerful and independent Kiev state, and justifies the rule of Kievan prince over all other Russian princes.

The direction set by Hilarion got reflected in subsequent annals, which made the basis for The Tale Of Bygone Years (beginning of the XII century), supposedly created by Nestor, a monk of the Kiev-Pechersk Monastery. In Sermon On Law And Grace, Hilarion sought to provide theoretical justification for independence of Russian state and strong princely power, while in The Tale Of Bygone Years, Nestor provided their historical justification.

Presenting the history of the Slavic tribes, describing the founding of Kiev and the emergence of Kievan state, the author of The Tale Of Bygone Years seeks to refute the Byzantine idea about the emergence of the Kievan state as a result of Christianization of Rus' under the influence of Byzantium. The annal states that Kievan princes are descendants of Varangian prince Rurik, who supposedly was urged by Slavs to manage them and establish order on the Russian land. The Russian statehood begins with Rurik, and his heirs, Kievan princes, by rights are senior among all the Russian princes.

These arguments of the annalist were used in XVIII-XIX centuries to create a "Norman theory" of origin of Russian state. Already in his time, M. V. Lomonosov has shown groundlessness of this theory. However, we must keep in mind that the author of The Tale of Bygone Years sought to refute the Byzantine idea of "lack of self-sustainability" of the Russian state with Varangian interpretation of "lack of self-sustainability." This solved immediate political tasks: claims of Byzantine for hegemony over Kievan Rus' got rejected, the significance of power of Kievan princes increased, emphasis was placed on their seniority, as well as inadmissibility of quarrels between princes under the conditions of the trend toward feudal fragmentation that seemed to be coming at the time. This incipient trend that emerged by XII century was slowed down by "Teachings" of Prince Vladimir Monomakh to his sons. In them, he instructed his sons to be wise princes, to continue the politics of strengthening and unification of Russian state, pacification of internal discord in it. Nevertheless, the period from the second half of the XII century to the first half of the XIII century (time preceding the Mongol invasion) is characterized by feudal fragmentation of Rus'. Calls and justification of an attempt to overcome crisis in Russian political ideology are reflected in The Tale Of Igor's Campaign, written by an unknown author after unsuccessful campaign of Novgorod-Severskiy Prince Igor Svyatoslavich against Polovtsy in 1185. The main patriotic idea of Sermon On Law And Grace is the idea of unity of Russian land, blood connection of all its parts, common interests in the face of external threat. The logic of unity of the Russian land that pervades The Tale Of Igor's Campaign was supposed to lead to and has led to, at the time historically justified, idea of undivided ruling, strong princely power.

In the end, the increase of power of grand princes was reflected in acceptance of title "autocrat of all the Rus'" by Ivan III, and then accession of the title "Tsar of all the Rus'" whose power is of divine origin ("we have an ordinance from God"). This idea was developed and completed in the theory "Moscow, The Third Rome," justified by Pskov monk Philotheus at the beginning of the XVI century. Moscow, the keeper of orthodoxy, became the Third Rome.

An interim summary is possible: synthesis of idea of Christianity and state power in the face of princes has led to strong, unlimited and, if necessary, brutal imperial power—a guarantee of greatness of the state, its prosperity. However, in the beginning there was destruction of the pagan mythological system by the Christian idea, which was first received in the face of a Grand prince, and then also the Tsar—performer-protector.

Monarchy And Philosophy

At the turn of XVII and XVIII centuries, Russian feudal state finalized as an absolute monarchy. The period of Peter's reforms was accompanied by a significant change in the ideological basis, content, and program positions of political and legal ideology. The official doctrine of absolutism included traditional theological justification for imperial power. Legislation (Military code, Spiritual regulations, and so on) stated: "The Emperor of All Russia is an autocratic and unrestricted monarch. God himself commands to obey his supreme power not only out of fear, but also for conscience." However, the main ways to justify Peter's reforms were references to the "common good." This idea was inherent in many absolute monarchies.

One of the first theorists of absolutism in Russia was a clergyman Theophan Prokopovich (1681-1736). In support of Peter's reforms and laws, Prokopovich wrote treatises, sermons, and other works. Prokopovich urged to combine "two earthly allowances—nature and science; and two given from above: catechistic knowledge of Christian faith and serious conviction in divinity of Holy Scripture."

The main politico-philosophical work by Prokopovich (The Truth Of Monarch's Will In Determination Of Heir Of His Powers...) was written in support of decree (1722) of Peter I, which determined the right of Emperor to appoint heir to the throne.

Vastness of territory became the official justification for autocracy. Theoretical search for an opposite class orientation has led to the idea of a federation, to ideas of representative democracy, to reasonable doubt in fatal predetermination of forms of state by geographical conditions. Finally, the idea of social contract, first used for an apologia for absolutism, was later on (and, with good reason) directed against autocracy.

Any thorough theoretical development in regard to the problem of limiting autocracy was prevented by extremely negative attitude of ruling monarchs towards ideas aimed at impairment of their power.

Ideology In Politics

Generation of new ideas begins when realization of the previous ends. There are some limitations. It has long been observed that "...teachings about rights and injustice are constantly challenged by a pen, as well as a sword, meanwhile teachings about lines and figures are not subject to dispute because the truth of the latter does not interfere with interests of people, does not collide neither with their ambition nor with their profits or desires. I have no doubt – wrote Hobbes, – that if the truth that three angles of a triangle are equal to two corners of a square contradicted someone's right to power or interests of those who already have power, then (since it would be at the mercy of those, whose interests are affected by this truth) teaching of geometry would be if not disputed, then supplanted by burning all books on geometry."

Considering some ideas, which formulate that the West has no authority over Russia, this can be taken at the golden mean: Russian national self-consciousness cannot be neither exclusively Slavophile nor exclusively

Western. The probability of Russian mission (to be a mediator between the East and the West) is negligibly small in modernity, but it was relevant at the time when the absolute monarchical power stopped satisfying Russian society ideologically. History of XIX and early XX centuries illustrates this process quite convincingly, as will be discussed below.

Georgi V. Plekhanov (1856-1918), one of the founders of the Russian Social-Democratic Labour Party, a prominent Marxist theoretician, like many other Marxists considered socialist revolution possible at that stage of development of capitalism, when material production reaches a high level and the proletariat will be composed of most of the population. He was a supporter of socialism imposed not by force, but by decision and with consent of the majority of people.

A prominent theorist of anarchism was Peter Kropotkin (1842-1921). Kropotkin expounded and justified the theory of anarcho-communism. He wrote that at the basis of society are equality for all, solidarity, and freedom. The originality of the concept is in that for some whole periods in history, according to Kropotkin, there was no state: the ancient republics of Greece, medieval free cities, free rural communities, in his opinion, were not states in the true sense of the word. "Throughout history of our civilization there are two currents, two opposing traditions: Roman and folk; imperial and federalist; tradition of power and tradition of freedom. And now, right before a great social revolution, these two traditions again stand face to face."

An independent outflow of Marxism is Bolshevism, the founder and leader of which was Vladimir Ilyich Ulyanov (Lenin) (1870-1924). The main idea of Bolshevism is the idea of a "party of a new type," a centralized organization of professional revolutionaries, seeking to become the ruling party, and realizing the Marxist program of building of socialism and communism.

In short, at the beginning of the XX century, diversity of political and legal doctrines increased, and that was generated by complication of social and political processes of social life, as well as multiplication of various methodological foundations of political and legal concepts that differed according to social-class basis. And, they were all focused on one goal: replacement of the old idea of absolute monarchy with some other, new idea. We all know well which one of the new ideas won. By the way, one of the first steps that the new authority made was replacement of all symbols that had roots in mythological images, and proceeded to declare the country atheistic. She was an excellent student of the old teachers who taught about the impossibility of preservation of old image-based structures, which can be equated with constantly forming additional structures that are in opposition to goals of the original structure, i.e. new authority.

Two Worlds: Real And Virtual

Not much time passed and the world began to change again. Realization of communist ideas was over. Time is required to generate new ones. However, there are no new ones yet. There was no time allotted to prepare, while Plekhanov, Kropotkin, Lenin, and many others had the time. Space was conquered practically without global battles, massive bloodshed, and other tantrums. The leap into informational society did not stipulate for exalted attributes. It was done impersonally, without ratings, with complete indifference to any attitudes towards it. The main thing is that it was and continues to be very beneficial in every way. However, problems began to surface. Firstly, existence of many worlds confuses the informational society, and on top of that a virtual world was added. The latter is so disarranged that according to words of "residents" of the virtual space themselves—it resembles an enormous informational trashcan. Secondly, the virtual world can be arranged, but a question arises: "How?" Since it is audio-visual, its space is limitless and unprecedented in its possibilities, but many of the components are not yet available. Thirdly, it is being rapidly mythologized. This is perhaps the most interesting point. In it, occurs a reverse process to that what we call history of legal and political doctrines. If all of these teachings had logic from an image to an idea, from religion to philosophy and methodology, then in the new world of informational society, where the virtual world takes up the central space—process from methodology to an image is clearly indicated.

Virtual space is not a new phenomenon. The world of spirits of a shaman, departure of Taoist to handwritten scroll, visions of saints of various confessions within themselves and in reality, and so on—in essence, all these are virtual worlds, but they are individual and not publicly available. Modern virtual world can be regarded as the embodiment of individual, supernatural worlds that are easily accessible by everyone without fasts, Great Schema, spiritual practice, where it is possible to get almost any divine or nearly divine quality as an ordinary user. There are no laws, politics are merely present, and there is no division into states and their representatives. There is law of a symbol, an image, art, knowledge, and culture. All this is led by ideology of creativity. And, if one of the main reasons for the large number, variety, and complexity of political and legal doctrines is the desire of each of the ideologues to defend the ideals of his class or his group and to refute ideology of the opposing class or group, then in the virtual world you are free to choose something and defend it if you want to or not; because by and large in this world no one needs you, and at the same time everyone needs you for your quality of production in this world. Requirements for quality are rising in proportion with the level of arrangement of the virtual world.

Conclusions

The conclusion is clear: we came back to that what was refused from the beginning. Thank God that myth turned out to have truly amazing survivability. From the time of rejection of myth to modernity, mythical symbol served as inspiration to many writers, poets, musicians, and artists. Dante, Shakespeare, Rabelais, A. Dürer, H. Bosch, E.T.A. Hoffmann, G. G. Byron, P. B. Shelley, M. Y. Lermontov, R. Wagner, F. Nietzsche, Joyce, T. Mann, C.G. Jung, S. Freud, V. Ivanov, F. Sologub, A. Borges, J. Amado, G. G. Márquez, and many, many others would likely have acceded to the opinion of Richard Wagner, who said that it is through myth that a nation becomes a creator of art, and that a myth is poetry of deep life views, which have a universal character. This becomes increasingly relevant in the emerging new world.

Bibliography

Meletinsky, E. M. (1963). *Proiskhozhdeniye geroicheskogo eposa* [On Origin Of Heroic Epos.]. Moscow: Izd-vo vostochnoy literatury.

Losev, A. F. (1965). *Stati i issledovaniya po yazykoznaniyu i klassicheskoy filologii* [Articles And Research Studies On Linguistics And Classical Philology.]. Moscow.

Likhachev, L. S. (1971). *Poetika drevnerusskoy literatury* [Poetics Of Ancient Russian Literature.]. Leningrad: Khudozh. lit.

Averintsev, S. S. (1972). *Analiticheskaya psikhologiya K.G.Yunga i zakonomernosti tvorcheskoy fantazii* [C. G. Jung's Analytical Psychology And Patterns Of Creative Imagination.]. Moscow: Iskusstvo.

Toporov, V. N. (1973). *O strukture romana Dostoyevskogo v svyazi s arkhaichnymi skhemami mifologicheskogo myshleniya* [On Structure Of Dostoevsky's Novel In Connection With Archaic Patterns Of Mythological Thought.].

Mints, Z. G. (1978). *O nekotorykh "neomifologicheskikh" tekstakh v tvorchestve russkikh simvolistov* [About Some Of The "Neo-Mythological" Texts In The Works Of Russian Symbolists.]. Leningrad.

Sovetskiy Entsiklopedicheskiy Slovar' (SES) [Soviet Encyclopedic Dictionary]. (1981). Moscow: Sovetskaya Entsiklopediya.

Losev, A. F., Meletinskii, E. M., Riftin, B. L., Toporov, V. N., Ivanov, V. V., Averintsev, S. S., ... Afanasyeva, V. K. (1982). S. A. Tokarev (Ed.). *Mify narodov mira* [Myths of the World] (Vols. 1-2.). Moscow: Sovetskaya Entsiklopediya.

Vygotsky, L. S. (1982). *Psikhologiya iskusstva* [Psychology Of Art.]. Moscow: Iskusstvo.

Otyugov, A. A. (1998). Corr. of RAO L. Verbitskaya & Assoc. Prof. B. Sokolova (Eds.), *Pervyy Rossiyskiy Filosofskiy Kongress: Chelovek – Filosofiya – Gumanizm* [First Russian Philosophical Congress. Human Being – Philosophy – Humanism]: *Vol. 7 Philosophy and Human Problem: Psychological Nature Of Mythological Consciousness*. St. Petersburg: SPSU Publishing House. ISBN 9785288018947. Presentation at the First Russian Philosophical Congress: Human Being – Philosophy – Humanism (in Russian), St. Petersburg. Retrieved from http://mirlyn.lib.umich.edu/Record/003947324 B4231.R6751997 (subscription required). [Also available at other libraries http://www.worldcat.org/title/chelovek-filosofiia-gumanizm-pervyi-rossiiskii-filosofskii-kongress/.]

Hamitov, I. (2002). *Filosofiya cheloveka: ot metafiziki k metaantropologii* [Philosophy Of A Human: From Metaphysics To Meta-Anthropology.]. Moscow: Nika-Tsentr.

Petrov, V. L. (2003). *Geopolitika Rossii* [Geopolitics Of Russia.]. Moscow: Veche.

ARTICLE 4

IDEOLOGY OF MONARCHY. FOR OFFICE OF THE HEAD OF THE RUSSIAN IMPERIAL HOUSE, HER IMPERIAL HIGHNESS GRAND DUCHESS MARIA VLADIMIROVNA.

Andrey Davydov

Translated by Kate Bazilevsky.

Office Of The Head Of The Russian Imperial House, Her Imperial Highness Grand Duchess Maria Vladimirovna 13 First Kozhukhovskiy Dr., Moscow, Russia, 115 280

Part 1. (Preliminary Theses)

EARTH

1. State is a territory defined by state borders, belongs to the whole nation in general and to every citizen in particular in that part, which is defined by ratio of one citizen to the total number.

2. The cost of property, defined as a part, is defined by ratio of the total value of extracted natural resources and the total number of citizens living in the Russian state.

3. A birth certificate or a passport of a Russian citizen is a lifelong action.

A birth certificate or a passport of a Russian citizen can be not only an identification card, but also a document that fixates the right of each person to a factual part of financial means from the development and sale of natural resources of Russia. Monthly reception of money by holders of birth certificate or passport of a Russian citizen from the extraction and sale of natural resources, regardless of geography of residence, nationality, religion, employment and occupation, can become one of foundations of the existence of Russia and its nation. Financial support of a birth certificate or passport of a Russian citizen by income derived from the development and sales of natural resources can act as a guarantor of equality of rights and sufficient financial base for women, women who intend to become mothers and those who have children, which can provide the primary basis for survival and health of the Russian nation going forward. The goal is equal basic financial conditions for each Russian citizen by birthright (as a Monarch). The banking sector can serve as an already existing mechanism for implementation. Thereby eliminating intermediation of state officials in the distribution of financial flows. This can remove a significant part of social tension, and allow every Russian citizen to voluntarily and on his or her own choose the sphere of application of his or her strengths, while the standard measure of this activity can be a figure of a Monarch.

NATION

1. A Russian citizen is the main value.

Existing social schemes are fundamentally divided into two halves. First: a human is part of a system (scheme). Second: a human is the basis and the end goal of the efforts of the system (scheme). The second half of the system (scheme) is used either formally or factually. A monarchy can be considered a third system (scheme), and a monarch as the total value of the highest (best) achievements of ethnic groups of his state. They can be the norm for the Monarch. This norm can be the etalon of qualities of personality for a particular person and for all the Russian people as a whole. In this case, specifically factual efforts of the socio-system are necessary, and the thesis about human achievement of the status of the primary, the main value of the state becomes fundamental.

2. The right to daily meals as the basis of biological existence of a human.

The right to life is fixed in the Constitution of Russia. Its protection is an obligation of state structures. The right to daily meals, as a basis for biological human existence, does not exist and remains a matter of personal concern. Financial support of a birth certificate or passport of a Russian citizen from sales of extracted natural resources can partially or completely eliminate this matter of personal concern and guarantee the right to life.

3. Culture is the foundation of the civilization, the extent of its survival and development.

Its essence is the totality of individuals who are carriers of research toolkit, knowledge, technological implementation, and experience. A cultured person is the embodiment of the image of a qualitative personality. The main direction is to know one's own self, i.e. a Human, the head of the genus, the only carrier and consumer of culture. Provisioning of informational supply is the basis of the development of culture. A personality that selectively closes the flow of information with the goal of self-cognition on itself—engages in informatization of the personality. With such direction of development of society it is possible to catalog informational space, informational society, its structuring in a hierarchical scheme. Metamorphoses from interaction between an individual and information can be structured in a finite variance. This will allow to control, regulate, and manage the informational society. This will also allow to shift the focus of the civilization from a human as a biological type, to subtype structuring of humanity. In such processes, a vector of direction towards a goal is necessary. Such a goal can be a modern cultural hero.

4. A modern cultural hero.

A modern cultural hero is a free informatized individual, equal to the society and able to participate in suppression, balance, and stimulation processes. Freedom of an individual is determined by the degree of his culture. This personality does not fit into the framework of the state, as it itself is a state that can exist without borders. Society is then considered as a ground for approbation of personality qualities of a citizen. With realization of the thesis of trinity of power, a confessional leader, a Monarch, a President, one of the tasks of intelligentsia can become the embodiment of the image of a cultural hero of modernity. In the cultural tradition of the civilization, this image is the bearer of certain qualities, which are part of the etalon personality—a Monarch. A description of the image of a cultural hero can become precisely that bearing point of cultural pillars, on basis of which can begin development of traditional values that automatically leads to creation of an ideology. In this process, a Monarch, as an etalon of personality, can become the goal of achievement, while the image of a cultural hero—an intermediate stage towards this goal.

5. Informatized individual and society.

Traditional values of the social community can be complemented by another one, which in time and space can become the main value—an informatized individual or society, consisting of informatized individuals, the main product of whom is a creative product. The state apparatus can sum up this product into an ideology and politics directed at ensuring work of the entire state, where the result of this activity must be totality of material and informational benefits for every citizen of the state.

EDUCATION

1. The goal of education is individuality of students with the division of the total number into categories in the general culture-forming process.

2. Continuation of school practice of acceptance and passing of examinations at higher education institutions—admission to all interested and examinations of those students who want to advance.

3. Equal opportunities for schoolchildren and students in the educational process provide the same standards for public schools and higher education institutions in the informational complex.

4. Standardization of rituals within the education system.

ECONOMY

1. Human-economy:

- a human is the main resource of the civilization
- the goal of economy is to make the economic scheme comfortable for use by a human
- a system of relations: human-human, human-machine (dictate of a human and transformation of a machine)
- rational use of a human and the economic scheme

In this scheme there is a possibility of evolutionary succession of generations in the established hierarchy of cultural values that can provide the optimal regime for expression of characteristic psychophysiological abilities of an individual. Break of this chain can lead to a holdup in evolution of the civilization.

2. Economy-human:

- a human is a secondary resource
- a system of relations: human-human, machine-human (dictate of a machine, stop of modernization of a machine, transformation of a human for a machine, and, as a consequence—general degradation)
- irrational use of a human and the economic scheme

LEADER

3. A Monarch is a living etalon of quality of human individuality for the Russian people. Spirit, spirituality is a Divine immaterial substance that provides natural functioning of the material. With natural development, material bio-forms have individual algorithms ranging from birth to death. Each bio-form, in the course of its unscathed development in the environment, acquires qualities, one of which is the natural beauty of the bio-form. When there are scathes due to this or that reason(s), a bio-form loses qualities that meet the concept of beauty and gets utilized by nature. Hence, it is evident that beauty as a concept characterizes the maximum degree of survival in space and time. A human is a bio-form that is able to universally exist in practically any natural environment, which is one of the fundamental differences between a human being and other bio-forms

(animals, plants, etc.). Another fundamental difference is inability to recognize in a particular person his natural program that provides natural individual development followed by productive activities based on it. Nevertheless, it is possible to identify psychophysiological parameters (factors): intellectual, physical, nutritional, emotional, sexual, and environmental, which interact with each other in space and time, where racial, ethnic, cultural and other indicators are correctors. A Monarch, being outside of correctors, takes care of interaction of the above-mentioned six factors and their quality, which determines the "place" of the Monarch, for example, by analogy with chess—outside the social chessboard. Qualities of the Monarch are the total value of the highest achievements of ethnic groups of his state, which is the norm for him. This norm can be the etalon of quality of an individual for the Russian people in general and for every citizen in particular. The technology to achieve such quality can be directed by formal and factual vectors. As a result, beauty can be expressed in the face of a Monarch through achievement of quality on six factors, and thereby embodying the path, the technology, the survival rate of the nation. This fundamentally differentiates a monarch from a figure that holds the highest political office (president, prime minister, etc.), as well as distinguishes a Monarch from highest confessional hierarchs, which determines the state's natural need for a monarch.

4. Power from birth (a monarchy) means responsibilities upon achievement of qualities of an individual used in the future political system, where the foundation is a citizen of the state. Monarchical power is power received from birth (financial dynasties are not considered here because they are a form of power received from "below"). It can be considered as a duty, which is intended to be carried out of inevitable necessity. A Monarch is an individual, realization of who's own desires is a secondary component. This is a fundamental difference from those who gain power "from below." Technologically, preparation of a Monarch goes from birth, where the principle "must do—must not do" is activated, rather than "I want—I do not want." He must know the natural development in time and space of all the state resources that he guards and ensure the normal course of all processes in the future. Based on stated above, it is obvious that power received from birth is an unavoidable duty to open up his abilities to the fullest, and these abilities will later on be used depending on dictates of time and space. In such an embodiment, a Monarch is the only acceptable natural third component to space and time, providing normal process of development of the state and its resources.

5. Monarch is a status that combines in itself qualities of spiritual and social leaders, and is a guarantee of wholeness and stability of processes.

A Monarch is the equivalent of supreme authority, connecting qualities of a spiritual and a social leader. Due to the lack of opportunities for spiritual and social leaders to maintain natural interaction in developing the legislative field of the state, a Monarch can be the third state power. With this existence there is an option to not have singular supreme spiritual and social authorities.

6. A Monarch is a preimage.

A Monarch is a preimage, which embodies the unity of the Russian ethnic diversity. A preimage is unchanging spiritual essence that gives the multiplicity of form. A preimage must correspond to (embody) the space, time and state of the nation. Space is the territory of Russia; time is the correlation of internal and external processes of society; state of the nation is the sum of cultural levels and welfare of the nation.

7. Confessional leaders—spiritual leaders of the Russian nation.

Confession is a spiritual tradition, where the main goal is to provide an individual path to God, according to individual characteristics defined by confessional hierarchs (representatives). The current situation is independent search by adepts for path to God within boundaries of a particular faith. Character, quality and criteria of manifestations of the Divine Spirit through the soul of adepts in the material world is the concern of religious workers. The Bible, the Koran, and other ancient books that underlie confessions—ritualize and regulate the human society, even though the definition of boundaries and areas of interaction between spiritual and social is not the prerogative of confessions. An attempt to use the ancient rules in modernity without changes can be considered anachronism. Changes can be done to the form, but not to the essence. If the essence changes, then state laws get modified or abolished and here already leader's personality acts as the criterion of legality. Change of form is possible through modern rite, ritual, rule, ceremony, etiquette, etc. Timely work in this direction means formation of new cultural references with presence of cultural heroes. Untimely, interrupted, unstable work on new forms leads to a lack of ideology, i.e. ideological vacuum, and in the future—to ideological chaos in the state; and as a consequence—to financial and political destabilization.

8. A President is the leader of society and regulator of social mechanisms.

Politics is the sphere of development of social interaction schemes. Schemes such as state policy, finance, manufacturing, trade, etc. are mechanisms, the ultimate goal of which is to lift all social problems off an individual in particular and the whole nation in general. Definition of boundaries and areas of interaction between the social and the spiritual is not the prerogative of a President—the political leader of society and regulator of social mechanisms.

9. Power achieved "from below" is a way of uncontrolled self-realization and realization of one's own personal desires, depending on social (career) level for a limited period.

Power is money. Money is not yet power. Power is achieved for the sake of money and uncontrolled, and, most importantly, unpunished realization of one's own obvious and hidden desires. The level of gained power determines degree of impunity. Existing shortcomings: space is used depending on desires, and time—irrationally. The first position—space with all the resources within it is not used and does not develop naturally. The second position—time gets spent on realization of desires in space, i.e. regardless of natural and necessary needs of the State. This automatically leads to social cataclysms that differ in their power and duration. This process can be infinite or, in modern conditions, considering the level of technogenesis of the civilization—finite.

ARMY AND ENFORCEMENT AGENCIES

1. The principle is "Caring about defeat."

All military doctrines are based on the principle "From defeat to victory and from victory to defeat." The principle of "Caring about defeat" calls for some differences from these doctrines.

In any military actions loss of material and human resources is foreseen, which allows us to consider these losses as part of the concept of "defeat," in spite of the resulting total or intermediate victory of one of the sides. The standard military expression that any war cannot be without loss does not make losses of human strength of the winning side into victory for fathers and mothers of the dead. To them this is grief, defeat, and loss of hopes, which were associated with their dead children. Economic damage resulting from military operations on the territory of the winner, which is inevitable, is not a positive (winning) article for the winning side. It also relates to the meaning of the word "defeat." The defeated side will seek revanche in the future, and that will lead to new military actions and inevitable defeat for one side, as well as the other. The principle of "Caring about defeats" first of all includes the goal of not having own losses of human strength and technology. This same principle allows to consider victory as a secondary result, which in the end stops the traditional swaying of the pendulum of military doctrines—"defeat-victory."

2. Modern state system of punishment (prison, camp, etc.)

The system of punishment (prison, camp, etc.) has two vectors. The first is the impossibility of further socialization of people after their release. The reason is the focus of prison and camp regimens towards acquisition of animal state by prisoners. Thus, places of imprisonment can be considered places for stable transformation of those people who pose a threat to the regime and state apparatus. The second is becoming aware of disadvantages of illegal actions and obtainment of multivariate possibility of integrating into the society or optimum realization of the individual after serving of the sentence. In the second vector, the thesis "Culture is the foundation of the civilization, the extent of its survival and development" is relevant. At its base, the system of punishments is coercive, which allows considering education on par with coercive labor. In connection with this statement, elements of standard norms of sets of places of imprisonment can be analogous to norms of public education institutions.

Prepared by Andrey Davydov,
acting deputy chief of Info-analytical administration.

Moscow (Russia)

January 2003

ARTICLE 5

ELITIST POLITICAL CONCEPTS

Andrey Davydov
Translated by Kate Bazilevsky.

Introduction

Nietzsche said that power does not know saturation, and unchangeably power that no one yet possesses, longs to be possessed and so ultimately this does occur.

The period from Plato and Aristotle to Marx, Engels, Lenin, Dunning, C. Merriam, A. Bentley, W. Lippmann, H. Laski, D. Gentile, G. Burdeau, G. Mosca, V. Pareto, H. Lasswell, K. Arrow, G. Becker, and many others can be considered a preparatory period in the cardinal modernization of society. And, in this case Russia cannot do with modernization of the political subject. For now the situation is saved only because neither political science nor psychology or other adjacent fields have no unified approach to the study of this problem.

Belongingness to the clan-oligarchic elite model is not an automatic pass onto the path of becoming such elite as a subject of development. Becoming clan-corporate structures as a subject of development in the modern development of civilization is difficult due to constraints of systemic nature. The division of the world system into producing information and consuming that information determines the limitations of possibilities of private political and financial subjects in resolving modernization problems of global (elitist) scale and actualizes strengthening of the role of an individual.

Consideration of the term "political elite" as a synonym of the concept of "political power" is largely an attempt to look at new values from the viewpoint of a political historian, who seeks analogues in the past. Unfortunately, today such excursus is unlikely to end in success because it is known that modern economic growth is distinguished by the primary role of scientific and technical progress, and intellectualization of main factors of production; intensity of Research and Development determines the level of economic development. According to predictions of experts, in the XXI century intellectualization of labor will become the main factor in global competition. Unfortunately, politics and politicians currently have only an indirect relation to such processes. They are more concerned with problems of power, state, and their own dislocation, as there is conviction in existence of political process as a perpetual motion machine. Unfortunately, it can only be concluded that as long as the elite takes care only of itself, as well as search, familiarization, distribution, and division of material resources—there cannot be any discussion about breakthrough into the world of the "select club" of intellectuals. In this club, the concept of "state" might already be nonexistent. And, one gets accepted there not based on merits or worth of personal capital, but based on the ability to meet requirements of a person of informational society of the "elite" class.

Fear Of Anarchism

In contrast to theories of state setup, anarchism advocated social revolution against the state. Anarchism rejects state power.

The first major theorist Proudhon understood abolition of all forms of human oppression, replacement of "political constitution" (beneficial only to ruling minority) with "social constitution" that corresponds to fairness and human nature. If transformation of formation of consciousness on the basis of synthesis of a human being and cybernetic systems as the most effective business is added to this, then it is already a revolution. And, very likely a revolution that is anarchistic.

The West is very afraid that Russians, ceasing to be Marxists, will not accept liberal democracy and will start to behave like Russians, rather than Westerners. In this case, the relationship between Russia and the West might again become hostile. The Russian elite traditionally did not understand the essence of ongoing process for the sake of the process itself, without a vector and without a goal, which is a "sinneth" of Western philosophical thought. The Russian version is a desire to find a pivot point, a vector, and a goal. And, in this case, the "old" anarchism in a new way might be claimed by Russia as a strategic goal. In this case, the West will not be able to claim a place for itself in the formula "state—informational society." In this formula, the concept of "state" does not fit in and becomes superfluous. In this context, Z. Brzezinski's "theory" about "American hegemony" over the Eurasian supercontinent seems quite jokey. American political scientist Francis Fukuyama seems overly ambitious as well. He put forward a concept of the "end of history," that the United States has already created the final, pragmatic model for the whole world, and that anything better is impossible, and therefore its common for all laws must become the "ethical" norm of life of united humanity. It is necessary to revisit the question of state in order to substantively examine the different variants of development of modern society.

State Is No Longer Power

Aristotle's thesis "state is power" is quite common now still (if one studies literature on this topic), but this definition is not the only one. From time to time formulations got reviewed simultaneously with processes when human communities changed slowly, but evolutionally. This process will continue until the end of existence of a human, who organizes himself in a more and more convenient society.

It is quite clear that power is money. However, money is not yet power. In modernity, a cybernetic human is an owner of values, which allow reorienting of society from the goal of having money to the goal of having (generating, producing, etc.) information, where money already plays a secondary function in countries that produce information, but not in countries-guilds.

Borders get erased with globalization processes, and that will ultimately result in their absence. Historically, an innovative type of development has formed in Western Europe and United States, and within the system of its factors the most important ones are economo-centric; economic interests of managing subjects coincide with interests of state, while the role of stimulants of development are fulfilled by internal economic needs. This might be the main miscalculation of the West and ungrounded concern of the less developed countries. Miscalculation consists in that with movement into informational society there cannot be separation into smart and stupid because the very dynamic of human development in the created situation implies that all people become smarter, not just select individuals. Solutions to economic issues in this process are in the role of tactical actions. It is clear that in the money arena there is an "iron law" of infinite capital appreciation and preferably without application of any effort, regardless of diligence and talent. In modern processes, this situation will be reviewed or is already being reviewed.

The Western world has emerged from hardships of '70s and '80s, and restored its role of global economic dominance, but a question arose: what to do next? The informational revolution, which allowed to link together regions of the planet, has helped. Cultural level of people who ensure work of communication systems became demanded, and that led to serious value of constantly growing cultural exchange in general, as well as between countries of the periphery and "first world." And, paradoxically, the success of functioning of the Western system, guaranteed by well-functioning mechanism, where everything spins in the right direction and rhythm, and defects get eliminated immediately—the requirements of Western society, with a known standard inherent in the apparatus of this society, no longer satisfy or suit members of this society.

Reorientation Of Elitism

An observed trend in reorientation from elitism of a subject to elitism of an object in general is obvious. Not the last role in this process plays the use of archetypal images in informational theory, where any class elitism disappears and in its place comes an overall elitism of personality.

Previously, religious, philosophical, ideological thought sought to justify the existence of elitism of a subject. Now, "modernity" is gradually expanding. "Modernity" in its essence is not new. Its commitment to rationalism and belligerent individualism, scientific knowledge, and human desire for comfort and well-being through production of material things, for continuous changes in the world through conquest of nature, for improvement of material tools of this conquest, conquest of new space, as well as mastership of new forms of human relations—today is quite justified.

Meanwhile, the Russian elite flounders under the impression that the weakening of state is a premise for democratization, contradicting ideas of modern political science, according to which democracy is possible only there, where an effective state still exists. Probably it is worth it to think about the existence of an effective state as an intermediate phase that allows sufficient time to transform our society by the global standards of the informational society. And, first of all, a qualitative transformation of elitist subject is necessary.

The Elite Of Modernity

The elite of our time are people who control, regulate, manage flows of information, through which a high-tech creative product goes from all and to all social layers without exception. The task of the modern elite is to make this process publicly accessible, and that will accelerate the process of transformations. Localization of this process and its closure onto the elite society is a task with no prospects.

Informational society cannot have classes because essentially there is nothing to divide. For example, in education there is no division into class of kindergartners, schoolchildren, and students. Apparently, symbolic denomination of phases is possible: a phase where intellectuals try to enter the informational society, intellectuals of this society, and etalon carriers of informational society. This process can be strict, similar to the evolutionary process.

In principle, people who by their qualities belong to the intellectual phase have the ability today to create finished products realistically without the use of labor of other people. Naturally, this might be connected with transformation of the process of formation of consciousness into an ultra-effective business, where self-programming of the controlling subject is possible in the face of national and transnational elites. It might be an interesting supposition that professional aspiration of PR-operators and their proprietors is to solve real-world problems by "brainwashing" with the use of a modern toolkit. This approach is effective to a limited extent, but if it begins to dominate, then that leads to a review of control systems.

High-performance of technology related to consciousness formation qualitatively raises the influence of political elites who have mastered it and those who use it. Such individuals and social groups become powerful. In this case, there is no "payment for power." Creation and implementation of new positions and values of orientation, a subject—an individual or a group (elitist)—forms a qualitatively new consciousness of society, considers himself a creator close to God. Euphoria of creativity gives him great satisfaction from everyday life.

Currently, it is possible to state that with all visible and formal successes of globalization (such as the informational revolution, the spread of financial and informational flows), it enters into the state of local crisis. This is manifested in the fact that the Western world has a claim from the rest of the world in regard to its active "self-closing," focusing on the scale of its part of the world. There is ground for this, as real statistics show that over the past decades there are very apparent tendencies in the global economy, which confirm that the Western world is becoming more and more independent from the rest of humanity, although it interacts with it more and more actively.

It must be stated that in some of the countries that are unable to fit into the informational economy, there is a sense that they are artificially deduced beyond its boundaries, become a superfluous element on the global scale. Revenue growth of rulers of corporations, incomes of all those who create their own companies in the high-tech field, incomes of programmers, researchers in these fields greatly exceed income of ordinary workers in the West, and this further contributes to states of their inferiority and worthlessness.

Today, developed countries demonstrate suppliers of qualitatively new limitless resource of information and knowledge, and in return get limited material resources from countries of the periphery. A new mechanism for formation of wealth has been created: in itself, the maximization of consumption in developed countries produces an increase in importance of human capital, which, unlike it is in Russia, has become their main resource. In connection with this, in the new historical situation it is necessary to attempt to reorient activities from creation of elite subjects to creation of an elitist society. This is possible because Russian political elites operate based on model of democracy of the West.

New Demographics

Human desire to belong to some class, and even more so—to the elite, is a normal desire. This occurs because, among other things, with any commonality of universal, interacting elements with a certain level of aggressiveness of the environment, there is a coming together into a structure capable of ensuring survival of quantity of elements of the structure. The world is changing. The aggressiveness of the environment is decreasing. And, elements of the structure, constantly improve connections between themselves in a direction of elimination of influence of the structure on these connections, and leave the structure upon achievement of disappearance of aggressiveness of the environment.

Every structure has taxonomy. Living and inanimate nature is already structured to a greater or lesser degree. The only "white" spot in the global taxonomy is a human himself. As for his own self, here he uses only the following parameters: skin color, which allowed racial identification, and division, according to Hippocrates, into choleric, sanguine, phlegmatic, melancholic. Unfortunately, latest studies in this field are not encouraging—the white spot has not been erased. It is possible to provide arguments as to why a catalog of human population does not exist today. There are various excuses in regard to this topic. For example, it is said that people are all original, individual, and unique. This is a widely accepted opinion, although physiologically we do not cardinally differ from each other. If we draw an analogy with a computer, then to imagine that each computer has individuality is a disaster for the computer world, as well as for the whole modern civilization. Quantitative excuses are also untenable. According to estimates by ornithologists: about 100 billion birds live on earth. This greatly exceeds the number of people, but it did not prevent their classification by a human. However, with appearance of the theory of information and understanding of his own self—a human—as a cybernetic system, the situation described above can be considered nonsense.

Research done by geneticists, particularly P. P. Gariaev, showed that at the cellular level, communication between cells is a packetized. According to P. P. Gariaev, this process is similar to videotape exchange, but it happens very fast. His experiments are aimed at clarification of the content of these tapes, as well as the recording method. The most successful trend in this area is synthesis of geneticists with followers of the theory of archetypes. An archetype acts as a subject of chimerical order. The form of an archetype consists of a variety of plant and animal segments, makes an organic whole, and is a holographic archetypal formula that determines cellular functionality, in other words—a human as a whole. It is very likely that cataloging of the human population can be considered through cataloging of individual archetypes.

Conclusions

In the era of globalization, management of the modernization process of a country presents itself to any elite as a problem of management of most important values, social, intellectual, psychological, and informational resources.

In the process of development and change of political and legal doctrines, historical destinies of three inherent structural elements (philosophical and methodological basis, theoretical content, and software requirements) differed.

Dynamics of development of scientific discoveries point in the favor of that in the near future it will be possible to connect the theory of information (informational individual, informational society, etc.) with genetics in the matter of cataloging of a human. Prediction of such situation is not part of this writing. Nevertheless, it should be noted that reception of the catalog, its approbation, and its use might lead to replacement of demography of quantity with demography of quality—the main criterion of the modern elitism.

Bibliography

Zenkevich, L. A. (Ed.). (1970). *Zhizn zhivotnykh* [Life Of Animals.]. Moscow: Prosveshcheniye.

Ushkov, A. M. (1995). *Sravnitel'naya politologiya: Zapad – Rossiya – Vostok* [Comparative Politics: West–Russia–East.]. Moscow.

Gariaev, P. P. (1997). *Volnovoy geneticheskiy kod* [Wave Genetic Code]. Moscow: Izdattsentr. ISBN 5781600221

Lobanov, V. V. (2001). *Analiz gosudarstvennoy politiki* [Analysis Of State Politics.]. Moscow: CUU.

Rakityansky, N. (2002) *Modernizatsiya Rossii i politicheskaya elita* [Modernization Of Russia And Political Elite.]. Vlast.

Danilevsky, N. Y. (2003). *Rossiya i Yevropa* [Russia And Europe.]. Moscow.

OUR OTHER BOOKS RELATED TO OUR SCIENTIFIC RESEARCH

Monographic Series

Archetypal Pattern. Fundamentals of Non-Traditional Psychoanalysis.

Davydov, A., & Skorbatyuk, O. (2014). K. Bazilevsky (Ed.). Anonymous (Trans.). *Archetypal Pattern. Fundamentals of Non-Traditional Psychoanalysis*: *Vol. 1. From Carl Gustav Jung's Archetypes of the Collective Unconscious to Individual Archetypal Patterns*. (Composed 2005. Original work published 2013 in Russian, ISBN 9781301447688.). San Diego, CA: HPA Press. ISBN 9781311820082

Davydov, A., & Skorbatyuk, O. (2014). K. Bazilevsky (Trans.). *Archetypal Pattern. Fundamentals of Non-Traditional Psychoanalysis*: *Vol. 2. Can Archetypal Images Contain Chimeras?* (Composed 2005. Original work published 2013 in Russian, ISBN 978130184859.). San Diego, CA: HPA Press. ISBN 9781310658570

Davydov, A., & Skorbatyuk, O. (2014). Arkhetipicheskiy Pattern. Osnovy Netraditsionnogo Psikhoanaliza [Archetypal Pattern. Fundamentals of Non-Traditional Psychoanalysis]: *Vol. 3. Archetype Semantics: How This Corresponds to the Concept of an 'Image'. How Archetypal Are Images?* (Composed 2005.). Marina Del Rey, CA: Catalog Of Human Souls GP. ISBN 9781301337309

Davydov, A., & Skorbatyuk, O. (2014). K. Bazilevsky (Trans.). *Archetypal Pattern. Fundamentals of Non-Traditional Psychoanalysis*: *Vol. 4. Society As A Community Of Manipulators And Their Subjects*. (Composed 2005. Original work published 2013 in Russian, ISBN 9781301399901.). San Diego, CA: HPA Press. ISBN 9781311809353

Catalog of Human Population - Non-Fiction Series

Individual (Subtype) Human Programs

Davydov, A., & Skorbatyuk, O. (2013). *Katalog Chelovecheskikh Dush: Programmnoye Obespecheniye Dushi Muzhchin/Zhenshchin, Rodivshikhsya <Data>* [Catalog of Human Souls: Software of Soul of Men/Women Born On <Date>] (Vols. 1-218. In Russian. Composed 2005-

2013.). Marina Del Rey, CA: Catalog Of Human Souls GP. [Available at http://www.humanpopulationacademy.org/pricing/ in all languages].

Human Manipulation Modes

Davydov, A., & Skorbatyuk, O. (2013-2014). *Katalog Chelovecheskikh Dush: Kak Podchinit' Muzhchin/Zhenshchin, Rozhdonnykh <Data>. Zhenskiy/Muzhskoy Manipulyativnyy Ctsenariy.* [Catalog of Human Souls: How To Subdue Men/Women Born On <Date>. Female/Male Manipulation Scenario.] (Vols. 1-39. In Russian. Composed 2005-2013.). Marina Del Rey, CA: Catalog Of Human Souls GP. [Available at http://www.humanpopulationacademy.org/pricing/ in all languages].

Ideologies

Davydov, A. (2014). K. Bazilevsky (Trans.). *Terrorism: A Concept For The ATC (The Commonwealth Of Independent States Anti-Terrorism Center).* (Composed 2001. Original work published 2014 in Russian, ISBN 9781311277848.). San Diego, CA: HPA Press. ISBN 9781310032189

Davydov, A. (2014). K. Bazilevsky (Trans.). *Ideology Of Monarchy. For Office Of The Head Of The Russian Imperial House, Her Imperial Highness Grand Duchess Maria Vladimirovna.* (Composed 2003. Original work published 2014 in Russian, ISBN 9781310150340.). San Diego, CA: HPA Press. ISBN 9781311970152

Davydov, A., & Skorbatyuk, O. (2014). K. Bazilevsky (Trans.). *Ideology Of Religions. Scientific Proof Of Existence Of "God": The Catalog Of Human Population.* (Original work published 2014 in Russian, ISBN 9781311946690.). San Diego, CA: HPA Press. ISBN 9781311413932 ISBN 9780988648593

Political Science

Davydov, A. (2014). K. Bazilevsky (Trans.). *Essence Of Political Ideologies And Their Role In The Historical Process (Political History Of Russia).* (Composed 2003. Original work published 2014 in Russian, ISBN 9781310199929.). San Diego, CA: HPA Press. ISBN 9781310199929

Davydov, A. (2014). K. Bazilevsky (Trans.). *Influence Of Psychophysiological Specifics Of A Leader On The Style Of Political Decision-Making.* (Composed 2003. Original work published 2014 in Russian, ISBN 9781310037832). San Diego, CA: HPA Press. ISBN 9781310104558

Davydov, A. (2014). K. Bazilevsky (Trans.). *Elitist Political Concepts*. (Composed 2005. Original work published 2014 in Russian, ISBN 9781310223228). San Diego, CA: HPA Press. ISBN 9781310822858

General Non-Fiction

Bazilevsky, K. (2012). *Human Population Academy: Laws of Human Nature Based on Shan Hai Jing Research Discoveries by A. Davydov & O. Skorbatyuk*. San Diego, CA: HPA Press. ISBN 9781301986781 ISBN 9780988648500

Davydov, A. (2013). *Shan Khay Tszin: Mify Ili Struktura Psikhiki?* [Shan Hai Jing: Myths Or Structure Of Psyche?] (Composed 1999. Originally pub. 1999 in Russian in Moscow: *Power Of Spirit*, 32-35.). Marina Del Rey, CA: Catalog Of Human Souls GP. ISBN 9781301590391

Davydov, A. (2013). *"Shan Khay Tszin" i "I Tszin" – Karta Psikhofiziologicheskoy Struktury Cheloveka?* [Shan Hai Jing and I Ching – Map of Human Psychophysiological Structure?] (Composed 2002.). Marina Del Rey, CA: Catalog Of Human Souls GP. ISBN 9781301510009

Davydov, A., & Skorbatyuk, O. (2014). K. Bazilevsky (Trans.). *AHNENERBE: Your Killer Is Under Your Skin* (Original work published 2014 in Russian, ISBN 9781311356741.). San Diego, CA: HPA Press. ISBN 9781311266682

A Man And A Woman – Non-Fiction Series

A Log With Legs Spread Wide

Davydov, A., & Skorbatyuk, O. (2014). K. Bazilevsky (Trans.). *A Log With Legs Spread Wide: Vol. 1. How Men Turn Women Into Nothing*. (Original work published 2014 in Russian, ISBN 9781310388125.). San Diego, CA: HPA Press. ISBN 9781311155771

Davydov, A., & Skorbatyuk, O. (2014). K. Bazilevsky (Trans.). *A Log With Legs Spread Wide: Vol. 2. How Goddesses Are Turned Into Logs. World History Of Turning Women Into Mats*. (Original work published 2014 in Russian, ISBN 9781311238894.). San Diego, CA: HPA Press. ISBN 9781311915603

Davydov, A., & Skorbatyuk, O. (2013). *A Log With Legs Spread Wide: Vol. 3. Women's Thirst For Power Over Men Is The Pathway To Become A*

Garbage. (Original work published 2013 in Russian, ISBN 9781301553075.). Marina Del Rey, CA: Catalog Of Human Souls GP. ISBN 9781301435500

Davydov, A., & Skorbatyuk, O. (2013). *A Log With Legs Spread Wide: Vol. 4. The Head – In The Underpants*. (Original work published 2013 in Russian, ISBN 9781301051281.). Marina Del Rey, CA: Catalog Of Human Souls GP.

Manipulative Games For Women

Davydov, A., & Skorbatyuk, O. (2013). *Manipulyativnyye Igry Dlya Zhenshchin* [Manipulative Games For Women]: *Vol. 1. March 23: Instruction for Exploitation of Men* (2nd ed., in Russian. Original work published 2005, Moscow: SNIALTotems. ISBN 9785716101333). Marina Del Rey, CA: Catalog Of Human Souls GP. ISBN 9781301803521

Davydov, A., & Skorbatyuk, O. (2013). *Manipulyativnyye Igry Dlya Zhenshchin* [Manipulative Games For Women]: *Vol. 2. April 6: Instruction for Exploitation of Men* (2nd ed., in Russian. Original work published 2005, Moscow: SNIALTotems. ISBN 9785716101302). Marina Del Rey, CA: Catalog Of Human Souls GP. ISBN 9781301069286

Davydov, A., & Skorbatyuk, O. (2013). *Manipulyativnyye Igry Dlya Zhenshchin* [Manipulative Games For Women]: *Vol. 3. October 13: Instruction for Exploitation of Men* (2nd ed., in Russian. Original work published 2005, Moscow: SNIALTotems. ISBN 9785716101326). Marina Del Rey, CA: Catalog Of Human Souls GP. ISBN 9781301900824

Davydov, A., & Skorbatyuk, O. (2013). *Manipulyativnyye Igry Dlya Zhenshchin* [Manipulative Games For Women]: *Vol. 4. December 7: Instruction for Exploitation of Men* (2nd ed., in Russian. Original work published 2005, Moscow: SNIALTotems. ISBN 9785716101319). Marina Del Rey, CA: Catalog Of Human Souls GP. ISBN 9781301413065

Secret Sexual Desires

Bazilevsky, K. (2013). *How To Seduce Men/Women Born On <Date> Or Secret Sexual Desires of 10 Million People: Demo From Shan Hai Jing Research Discoveries by A. Davydov & O. Skorbatyuk*. (Vols. 1-10). San Diego, CA: HPA Press.

Bazilevsky, K. (2013). *How To Seduce Men & Women Born On March 5 Or Secret Sexual Desires of 20 Million People: Demo From Shan Hai Jing Research Discoveries by A. Davydov & O. Skorbatyuk*. San Diego, CA: HPA Press. ISBN 9781301087204

Bazilevsky, K. (2013). *Secret Sexual Desires of 100 Million People—Seduction Recipes For Men & Women: Demos From Shan Hai Jing Research Discoveries by A. Davydov & O. Skorbatyuk*. San Diego, CA: HPA Press. ISBN 9780988648579 ISBN 9781301135035 ISBN 9780988648586

A list of other publications related to our scientific research can be found at http://www.humanpopulationacademy.org/publications/.

CONNECT WITH US

1. Visit our official website.

Human Population Academy and Special Scientific Info-Analytical Laboratory—Catalog of Human Souls:
http://www.HumanPopulationAcademy.org

2. Connect with us on social networks.

- *Facebook* - http://www.facebook.com/HumanPopulationAcademy (Note: you must be logged in to *Facebook* in order to access this page.)
- *YouTube* - http://www.youtube.com/user/HumanPopulAcademy
- *Google+* - http://plus.google.com/+HumanpopulationacademyOrghumannature
- *LinkedIn* - http://www.linkedin.com/company/2484433
- *Pinterest* - http://pinterest.com/humanpopacademy/
- *Twitter* - http://twitter.com/HumanPopAcademy

3. Contact us.

You can find out how to contact us at the Human Population Academy's website under Contacts (see https://www.humanpopulationacademy.org/breakthrough-discovery/contacts/).

ABOUT US

Special Scientific Info-Analytical Laboratory—Catalog of Human Souls was founded by Andrey Davydov. The laboratory is engaged in research and decryption of the ancient Chinese monument Shan Hai Jing, as well as other ancient texts, and creation of the *Catalog of Human Population*. The technology of uncovering individual structures of psyche of *Homo sapiens* for this Catalog was developed by Andrey Davydov; it is not based on any existing domestic or foreign research, methods or theoretical concepts. The laboratory is a partner with the Human Population Academy.

Human Population Academy was founded by Kate Bazilevsky. The Academy‘s mission is to inform all of over 7 billion humans living on Earth about the discovery of the *Catalog of Human Population*. The Academy educates about the *Catalog of Human Population* (*Catalog of Human Souls*) and provides access to informational materials from this Catalog to the public.

LEADERSHIP

ANDREY DAVYDOV

Research Supervisor of the Special Scientific Info-Analytical Laboratory—Catalog of Human Souls

Andrey Davydov is an expert in Chinese culture, researcher of ancient texts, the author of scientific discovery of the *Catalog of Human Population* and the technology of decryption of the ancient Chinese monument Shan Hai

Jing as the *Catalog of Human Population*. He authored over 300 published books, including scientific monographs and ideologies. In 2012, he was granted political asylum in the USA due to persecution by a group of employees of the Federal Security Service of Russian Federation (FSB, formerly KGB), who decided to expropriate his research product—the *Catalog of Human Population*.

OLGA SKORBATYUK

Senior Analyst at the Special Scientific Info-Analytical Laboratory—Catalog of Human Souls

Olga Skorbatyuk is a professional psychologist, one of the developers of the *Catalog of Human Population*, the founder of Non-Traditional Psychoanalysis, and co-author of over 300 books and scientific articles. She was granted political asylum in the USA together with A. Davydov.

KATE BAZILEVSKY

Founder of the Human Population Academy, Junior Analyst at the Special Scientific Info-Analytical Laboratory—Catalog of Human Souls

Kate Bazilevsky is the director of the Human Population Academy, a Junior Analyst at the Catalog of Human Souls laboratory, an author and a translator of books about the *Catalog of Human Population.* She holds a degree in MIS and psychology. She founded the Human Population Academy in 2011 and a publishing company called HPA Press in 2012.

www.ingramcontent.com/pod-product-compliance
Lightning Source LLC
LaVergne TN
LVHW022055200926
848481LV00003B/914